AF793961

Autonomic Nervous System Monitoring - Heart Rate Variability

Edited by Theodoros Aslanidis

Published in London, United Kingdom

IntechOpen

Supporting open minds since 2005

Autonomic Nervous System Monitoring - Heart Rate Variability
http://dx.doi.org/10.5772/intechopen.77922
Edited by Theodoros Aslanidis

Contributors
Adam Potter, David Looney, Xiaojiang Xu, William Santee, Shankar Srinivasan, Jonathan V Roth, Robert Drury, Julian Torres-Jacome, Alondra Albardo-Ibañez, Rosa Elena Arroyo-Carmona, Daniela Alexandra Bernabe-Sanchez, Marissa Limón-Cantú, Benjamin López-Silva, Martha Lucia Ita-Amador, Moacir Fernandes Godoy, Michele Gregório, Giovanni Minarini, Noor Aimie-Salleh, Nurhafiezah Hasanudin, Siti Nur Shakiroh, Nurul Aliaa Abdul Ghani, Yousif Mohamed Y. Mohamed Yousif Abdallah, Nouf Abuhadi

© The Editor(s) and the Author(s) 2020
The rights of the editor(s) and the author(s) have been asserted in accordance with the Copyright, Designs and Patents Act 1988. All rights to the book as a whole are reserved by INTECHOPEN LIMITED. The book as a whole (compilation) cannot be reproduced, distributed or used for commercial or non-commercial purposes without INTECHOPEN LIMITED's written permission. Enquiries concerning the use of the book should be directed to INTECHOPEN LIMITED rights and permissions department (permissions@intechopen.com).
Violations are liable to prosecution under the governing Copyright Law.

Individual chapters of this publication are distributed under the terms of the Creative Commons Attribution 3.0 Unported License which permits commercial use, distribution and reproduction of the individual chapters, provided the original author(s) and source publication are appropriately acknowledged. If so indicated, certain images may not be included under the Creative Commons license. In such cases users will need to obtain permission from the license holder to reproduce the material. More details and guidelines concerning content reuse and adaptation can be found at http://www.intechopen.com/copyright-policy.html.

Notice
Statements and opinions expressed in the chapters are these of the individual contributors and not necessarily those of the editors or publisher. No responsibility is accepted for the accuracy of information contained in the published chapters. The publisher assumes no responsibility for any damage or injury to persons or property arising out of the use of any materials, instructions, methods or ideas contained in the book.

First published in London, United Kingdom, 2020 by IntechOpen
IntechOpen is the global imprint of INTECHOPEN LIMITED, registered in England and Wales, registration number: 11086078, 7th floor, 10 Lower Thames Street, London, EC3R 6AF, United Kingdom
Printed in Croatia

British Library Cataloguing-in-Publication Data
A catalogue record for this book is available from the British Library

Additional hard and PDF copies can be obtained from orders@intechopen.com

Autonomic Nervous System Monitoring - Heart Rate Variability
Edited by Theodoros Aslanidis
p. cm.
Print ISBN 978-1-83880-518-0
Online ISBN 978-1-83880-519-7
eBook (PDF) ISBN 978-1-83880-577-7

For EU product safety concerns:
IN TECH d.o.o., Prolaz Marije Krucifikse Kozulić 3, 51000 Rijeka, Croatia, info@intechopen.com or visit our website at intechopen.com.

We are IntechOpen,
the world's leading publisher of
Open Access books

Built by scientists, for scientists

4,800+
Open access books available

122,000+
International authors and editors

135M+
Downloads

151
Countries delivered to

Our authors are among the
Top 1%
most cited scientists

12.2%
Contributors from top 500 universities

CLARIVATE ANALYTICS BOOK CITATION INDEX INDEXED

WEB OF SCIENCE™

Selection of our books indexed in the Book Citation Index
in Web of Science™ Core Collection (BKCI)

Interested in publishing with us?
Contact book.department@intechopen.com

Numbers displayed above are based on latest data collected.
For more information visit www.intechopen.com

Meet the editor

Dr. Theodoros K. Aslanidis received an MD from the Medical University Plovdiv, Bulgaria, and a PhD from Aristotle Medical University of Thessaloniki, Greece. He served in the Hellenic Army Force as a medical doctor and then worked as a rural physician at Outhealth Centre, Iraklia and Serres' General Hospital, Greece. He completed his residency in anaesthesiology at Hippokration General Hospital, followed by training in Critical Care at AHEPA University Hospital and a postgraduate program in Prehospital Emergency Medicine. Dr. Aslanidis served as an EMS Physician and Emergency Communication Center Medic at the National Centre of Emergency Care. Currently he is a consultant-researcher at the Intensive Care Unit of St. Paul General Hospital, Thessaloniki, Greece. He serves as editor for multiple medical journals. His research interests include medical writing, data analysis, emergency critical care, and neuromonitoring.

Contents

Preface

The autonomic nervous system (ANS) plays a vital role not only for homeostasis of the organism but also for reaction rate and extent to stimuli. Thus, monitoring ANS is crucial for understanding disease progress. There are many methods for monitoring ANS, including heart rate variability (HRV), hormonal biomarkers (epinephrine, cortisol, norepinephrine), infrared digital pupillometry, salivary α-amylase, electrodermal activity recording, and functional nervous system imaging.

Every ANS marker is considered a complex reflection of the sympathetic—parasympathetic system balance activation (autonomic outflow), neuroendocrine influences, and the ability of the human system network to respond to the former factors (autonomic responsiveness). They are probably more than an indicator for probable disturbances in the autonomous system. They can serve as a surrogate indexes both for the objective well-being (homeostasis and health status) and for subjective well-being (emotional and psychological health). Increasing data support in its use of ANS for monitoring for both somatic and psychological disorders and diseases. Moreover, newer studies provide us an insight into the physiology of consciousness and raise our understanding of several psychological and physiological processes, like neuroendocrine habituation or emotional regulation.

This book focuses on HRV, which refers to the interval between R waves in the electrocardiogram (i.e., variability in beat-by-beat heart period). The first section of the book is dedicated to technical themes related to both modes of monitoring and the variables recorded. The second section highlights special aspect use of heart rate variability HRV in relation to hypothermia. Finally, the third section of the book covers general aspects of its HRV application. Throughout the book, the authors offer us not only a "vigorous" review of the current literature but also a research road path for further advancement.

Theodoros Aslanidis, MD, PhD
Consultant/Researcher,
Anaesthesiology - Critical Care,
Pre-hospital Emergency Medicine,
Intensive Care Unit,
St. Paul General Hospital,
Thessaloniki, Greece

Section 1

Technical Themes

Chapter 1

HRV in an Integrated Hardware/ Software System Using Artificial Intelligence to Provide Assessment, Intervention and Performance Optimization

Robert L. Drury

Abstract

Heart rate variability (HRV) is increasingly recognized as a central variable of interest in health maintenance, disease prevention and performance optimization. It is also a sensitive biomarker of health status, disease presence and functional abilities, acquiring and processing high fidelity inter beat interval data, along with other psychophysiological parameters that can assist in clinical assessment and intervention, population health studies/digital epidemiology and positive performance optimization. We describe a system using high-throughput artificial intelligence based on the KUBIOS platform to combine time, frequency and nonlinear data domains acquired by wearable or implanted biosensors to guide in clinical assessment, decision support and intervention, population health monitoring and individual self-regulation and performance enhancement, including the use of HRV biofeedback. This approach follows the iP4 health model which emphasizes an integral, personalized, predictive, preventive and participatory approach to human health and well-being. It therefore includes psychological, biological, genomic, sociocultural, evolutionary and spiritual variables as mutually interactive elements in embodying complex systems adaptation.

Keywords: heart rate variability, HRV, health, well-being, health biomarker, high fidelity data acquisition, digital epidemiology, KUBIOS platform, high-throughput artificial intelligence, implantable biosensors, iP4 health model, complexity theory, complex adaptive systems

1. Introduction

The history of science shows clearly that the development of new techniques and tools of observation lead to improved scientific understanding and the development of more adequate theories. The development of the telescope and microscope made for conceptual breakthroughs in both the physical and biological sciences, facilitating empirical observations that allowed astute observers to create new and more powerful conceptual approaches.

This chapter will describe the development of heart rate variability (HRV) as a meaningful variable to monitor, interrogate and intervene in the functioning of the human nervous system and the psychophysiological systems it communicates with [1, 2]. It will provide a context for understanding HRV's role in the assessment, maintenance and enhancement of human health, well-being and performance. Further, appropriate techniques for studying HRV will be explored and a variety of applications will be described. Finally, iP4, a systems-based model of human health and optimal performance will be described, and several HRV-based integrated hardware/software systems will be described that exemplify that model.

2. HRV and the nervous system

Since the development of the original neuron approximately 600 billion years ago in worm-like creatures, the nervous system has emerged into increasingly complex and multifunctional neural networks that have vastly increased the adaptive capabilities of those organisms so endowed. This typifies the evolutionary process that has been studied as complexity theory within the systems view life [3]. Recently, the understanding of complex adaptive systems has been aided by the application of non-linear systems dynamics, as a supplement to more traditional linear modes of exploration and understanding. Increasingly complex entities emerge through processes of self-organization in interaction with environments demanding fitness to form adaptive systems which consist of multiple interactive and interdependent coevolving components. In the case of humans, the nervous system has played a decisive role in the increasingly dominant position currently occupied in the planetary ecosystem.

The central nervous system (CNS), composed of the brain and spinal cord has been historically identified as the most important part of the nervous system for conventional scientific study [4], with Kandel devoting only one chapter out of 64 to the autonomic nervous system (ANS). It is becoming increasingly clear that the peripheral nervous system (PNS) plays a crucial part in the remarkable abilities of humans. In particular, the ANS is known to mediate the sophisticated homeostatic dynamics that allow organisms to maintain a relatively stable interior environment needed to carryout complex adaptive tasks and supports the affective elements that comprise the significant motivational features characteristic of humans. The two major subdivisions of the ANS are the sympathetic and parasympathetic nervous systems [5]. The sympathetic nervous system is associated with energizing the organism during times of threat or challenge. Such activities have been described as "fight or flight" responses. The parasympathetic nervous system has been found to exert calming, stabilizing or reparative effects described as "tend and befriend" responses. A key structural and functional component which modulates the dynamic homeostatic balance is the vagus nerve complex, which originates in the brain stem and is widely connected with major organs such as the heart, lungs, stomach, genitals, pharynx, larynx, facial musculature, and middle ear muscles [1]. In addition to the stress "fight or flight" and calming "tend and befriend" responses, the vagus mediates the equally important "freeze" or immobilization response which is associated with death feigning in many species possessing the vagal nerve complex. These three response elements are integral to HRV, which is defined as the amount of variance in R-R wave intervals, also called the interbeat interval (IBI). The IBI is used to calculate the moment by moment variations in heart rate which constitutes HRV.

In addition to the systems described above, other neural network systems play significant roles in the overall integrated functioning of the human organism.

A key example is the enteric nervous system (ENS) which is an integral part of the enteric region and bidirectionally communicates with the CNS and ANS [6]. It has been described by some investigators as a "second brain" and functions largely independently, generating significant amounts of the neurotransmitter serotonin, important in proper neurobehavioral functioning. It also may be a key participant in the functioning of the immune system and mediate the role of the enteric microbiome. It is proposed here that the increasing empirical understanding of the vagal nerve complex warrants the use of the term cardio-vagal nervous system (CVNS). Notably, the role of the vagus nerve and its role in determining HRV, which has been studied for over 150 years [7], have been expanding rapidly with over 26,000 citations resulting from a recent PubMed search of the terms "HRV" and "heart rate variability."

3. Heart rate variability and the vagal nerve complex

The term vagus derives from the Latin term for "traveler." As described above, this is apt as the vagal nerve complex is widely distributed throughout the body. Its origins in the subcortical region of the CNS have been identified by Porges [1] as the nucleus ambiguus, the dorsomedial medulla and the nucleus tractus solitarius. These three neural structures represent the primary central regulatory components of the vagal complex and are responsible for three significant functions. The sympathetic mobilization for fight or flight has been recognized for some time, while the dorsal vagal response is a vestigial immobilization/death feigning system and the ventral vagal complex mediates the social engagement system for adaptive motion, emotion and communication. This is perhaps its most important feature to social organisms such as humans, where communication and mutual support have been identified as crucial aspects of evolutionary fitness, contributing to both biological and cultural evolution. This conceptual approach has been called the polyvagal theory by Porges [1] and the neuro-visceral integration model by Thayer and Lane [2].

While these functions are not currently subject to isomorphic assessment, it has been demonstrated empirically that HRV is an accurate and sensitive measure of the actions of these three subsystems. HRV is defined as the instantaneous variability found when continuous R-R intervals in the EKG are recorded [8]. These intervals are easily recorded using both standard 12 lead EKG protocols and a wide variety of freestanding equipment whose quality ranges from adequate to poor. It is impossible to obtain reliable HRV data when the equipment used to acquire the R-R intervals is either lacking reliability or is poor fidelity. If high fidelity quality data are obtained, there are three primary approaches to data analysis that have been found valuable: frequency domain measures, time domain measures and nonlinear measures [9]. Each of these approaches have demonstrated utility, although it has also been shown that some measures are less sensitive to salient phenomena and therefore, selection of the most robust analytic strategy is an important area of ongoing investigation and will be discussed further in the section of algorithmic analysis, artificial intelligence and related issues.

The research literature on HRV indicates that it can be a sensitive biomarker for a wide variety of disorders and conditions [10]. This includes medical disorders such as all-cause mortality, sudden cardiac death, sepsis, myocardial infarction, diabetic neuropathy, transplantation issues, myocardial dysfunction/heart failure and noncardiovascular diseases such as Alzheimer's dementia, epilepsy, diabetes, tetraplegia and liver cirrhosis [11]. It is important to note that in some conditions such as sepsis, the onset of subjective symptoms is often delayed and makes effective intervention difficult or impossible. However, HRV is frequently suppressed before these subjective reports occur, giving a crucial advanced warning of serious developments. HRV has also been shown to be sensitive to psychosocial disorders and dysfunctions such as depression, anxiety, bipolar disorder, attention deficit/hyperactivity disorder, substance abuse/craving disorder and post-traumatic stress disorder. HRV has also been used as a sensitive monitoring strategy in pacing physical training and determining when rest and recovery are indicated to avoid overtraining. Similarly, HRV has been shown to be an indicator of the level of executive functioning and resilience, both positive psychological phenomena. Another positive adaptive measure is the respiratory sinus arrhythmia (RSA), which is observed when HR increases during inhalation and decreases during exhalation. Notably, when a person is near death, the RSA, and therefore, HRV, are diminished or nonexistent. It should be noted that while HRV fulfills the epidemiological virtue of relatively high sensitivity, it does not possess high specificity, and therefore a careful consideration of contextual factors is necessary to make HRV a useful biomarker of health [12, 13]. In general, reduced HRV indicates impairment or dysfunction, while increased HRV shows improved functional or health status.

4. Methodological and technical issues

Accompanying the explosive growth of interest and research on HRV (12) have come a number of significant issues and problems that can impede progress. While the standard 12 lead EKG protocol is widely used in conventional medicine to produce high quality data, it is cumbersome, obtrusive and expensive. It also lacks the necessary data analytic software capable of recording and interpreting multiple HRV domains. These issues limit accessibility and more widespread use. A number of devices are available for research and clinical applications, some of which use hardwired photoplethysmography and occasionally wireless photoplethysmography accomplished by Bluetooth, relieving the individual from being physically connected to the equipment. Research studies have shown that implanted sensors may acquire interbeat interval data from which HRV can be derived. The interbeat interval data is either stored for later analysis or processed onboard with some type of feedback "HRV" score generated. The actual details regarding the meaning of some composite "HRV" scores is not always clear.

Similar devices are offered on the consumer market, targeting customers wishing to enhance or "fine tune" their physical training regimens. These devices most often use either ring or watch based sensors to obtain interbeat data and reliability data are not generally available, but it is likely that these modes of data collection and analysis contain significant artifacts and other data flaws more accurate chest straps and photoplethysmography sensors are seldom used. A transparent approach to the operation of such devices is desirable to examine both the reliability and validity of such devices, although manufacturers sometimes proclaim proprietary interests which shield them from this type of accountability. A technically feasible approach for interbeat data collection is to use implanted sensor systems which combine high fidelity data acquisition with Bluetooth data transmission and wireless capacitive battery charging so that the device could remain in place for a significant period, even indefinitely. This method would allow collection of more longitudinal data and make data collection in the natural environment relatively simple. Such data would be invaluable in determining a person's baseline state, a feature that is often missing in brief "snapshot" assessments. With such baseline data, a much more meaningful "vital sign" could be available, not only during office visits, but at any other time deemed relevant. It would also popularize the use of

HRV, since one of the drawbacks cited by many individuals is the cumbersome and inconvenient nature of currently available devices. Using the Bluetooth protocol to transmit data to a smart phone would also make data collection and analysis much more accessible, since smart phones are very widely used globally, including in areas with no other communication resources. With appropriate machine learning algorithms, the smart phone could also mediate actionable patient prompting or intervention, including HRV biofeedback. Using the digital epidemiological approach of population surveillance, ongoing monitoring could be available to distant healthcare facilities to prompt more detailed assessment and/or intervention at the individual or group level. Since this approach could involve an implanted device and longitudinal assessment, it might be perceived as more "invasive" and raise privacy and confidentiality issues, especially as used in healthcare contexts. Such concerns are legitimate and would need to be carefully addressed, although data collected by other means is equally worthy of such consideration, especially at this time when individual's data are regularly "harvested" or "scraped" surreptitiously by commercial ventures and monetized.

Until our understanding of HRV improves through machine learning and other artificial intelligence approaches, any one metric amongst the more than a dozen available, is somewhat incomplete. Currently, the SDNN time domain measure (standard deviation of interbeat intervals) is most frequently used and has value. A very comprehensive systemic software suite has been developed by Tarvainen and colleagues [14] called KUBIOS. It is available for analysis of HRV in multiple modes of time domain, frequency domain and nonlinear modes using a batching approach. A current limitation, however, is that KUBIOS does not conduct its analyses in real time, but that limitation is being addressed and a real time version of KUBIOS is in development for purposes of both scientific clarity and consumer use [15]. The KUBIOS platform offers many benefits such as multi-method analytic strategies and clear documentation, making it ideal for increasingly popular big data approaches such as artificial intelligence, machine learning, and high-throughput and cloud computing. Such approaches are especially applicable to the large number of data points that can be collected in longitudinal HR data.

5. Innovative applications of HRV

Given the popularity and "sizzle" of developments in the big data areas of artificial intelligence, machine learning and high-throughput cloud computing, a recent report by Liu and colleagues [16] illustrates an area of high potential value. Building on the work of King and associates [17] on the use of HRV in decision support for trauma patients being transported by helicopter to a trauma center, Liu used machine learning to create predictive models that could detect the need for lifesaving interventions. Their results were near perfect predictions with receiver operating characteristics under the curve = 0.99. While their preliminary report suffered from issues such a relatively small sample size and difficulty extracting high fidelity HRV data, it is still proof of concept that such a systems approach is feasible and relevant.

Another important model described by the National Institute of Standards and Technology as the Analysis as a Service (AaaS) has been developed and exemplified by IBM's Watson Analytics (WA), which is a cloud based AaaS that claims to "carry out a number of significant data analysis and display approaches in a user friendly manner" [18]. The Explore and Predict modalities use a variety of data clustering and machine learning approaches that can go far beyond single variable linear prediction, while the Assemble modality develops effective data display and infographics.

The use of WA by Guidi and colleagues [19] demonstrates proof of concept for a cloud-based data acquisition and analysis system which can make accurate clinical diagnostic decisions differentiating patients with heart failure from normal individuals on the basis of HRV. The process developed by Guidi involves data acquisition using the PhysioBank and PhysioNet to obtain and categorize ECG data into the appropriate format of R to R intervals using the PhysioNet HRV Toolkit [20]. This data consisted of 15 subjects with severe heart failure, 29 subjects with moderate heart failure and 54 healthy subjects with normal respiratory sinus arrhythmia. All subject data was initially collected using standard ECG protocols. The resulting data set was examined by WA and a variety of commonly used HRV statistics were derived. These statistics were compared to the data available in the current literature. This shows the results concerning accuracy of prediction using the Total Power HRV (TOT_PWR) statistic with a 90% predictive accuracy.

The use of such tools in critical care is exemplary and similar approaches have been suggested by Drury [21] in the areas of more routine clinical care and digital epidemiology. As noted by Topol [22], while the use of the terms artificial intelligence and machine learning has tended to be overblown, the use of existing big data sets and extensive longitudinal data as proposed here is ideal for such an approach and can assist in the laborious and sometimes obscure task of empirical investigation. I have previously suggested [10] using a wireless implantable high fidelity data acquisition device, networked by Bluetooth to a suitable machine learning version of the functionalities possessed by KUBIOS could interrogate the data sets for the creation of the most suitable predictive models for making not only clinical decisions based on changes in patient status, but monitor patients regarding the emergence of new conditions. This type of monitoring could also realize the concept of digital epidemiology, including the use of self-monitoring prompts to aid individuals in appropriate help seeking and self-regulation strategies. See **Figures 1** and **2** to visualize the use of such a system and its ability to discriminate rest, activity and recovery periods [23]. As is indicated by the development of groups such as the quantitative self and biohackers, there is a social demand to assist individuals in optimizing their health and well-being and enhancing their performance in a wide variety of areas. As proposed here, these groups have sometimes used

Figure 1.
ReThink wireless biosensor with capability to acquire HR, three-dimension accelerometry and respiration data. These data are then transferred to a Bluetooth enhanced personal device, as well as stored there. Quarter is shown for size comparison, with dimensions of 7.5 mm width, 4 mm height, and 1.4 mm depth (used with permission).

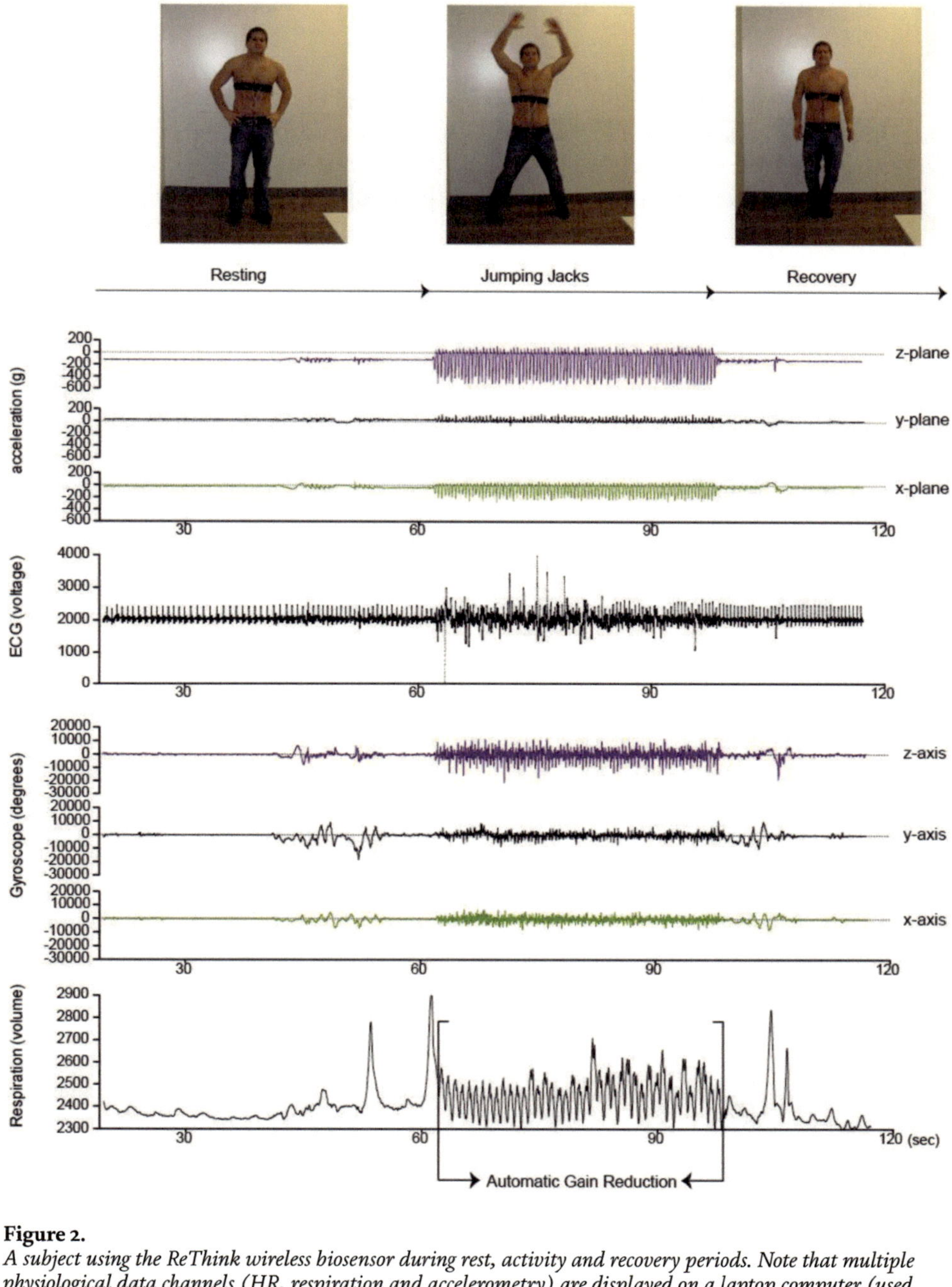

Figure 2.
A subject using the ReThink wireless biosensor during rest, activity and recovery periods. Note that multiple physiological data channels (HR, respiration and accelerometry) are displayed on a laptop computer (used with permission).

implanted devices to monitor their physiological status. The professional community has perennially spoken through the Institute of Medicine, the National Science Foundation, the National Institutes of Health, the World Health Organization and other organizations of the need for accessible, safe and effective healthcare.

The development and utilization of predictive models derived from artificial intelligence approaches such as machine learning could be beneficial in many aspects of care provided by the current healthcare industry. If clinicians had highly specific empirically based decision support data readily available, their interactions with patients might be more timely, less stressful and "more human," to use Topol's phrase, with less time with the care provider on the computer and more face to face interaction. If ongoing HRV monitoring was used to track clinical status of patients, the need for routine exams, especially of the "worried well" would likely decrease. Prompt intervention for patients whose ongoing status was being monitored would be more likely, thus addressing the serious problems of both under and overutilization of medical services. This would accomplish a transition away from the expensive intensive care model to a less expensive, more extensive model. As is the case with the work of Liu's proof of concept, this approach is currently feasible and the elements of such a hardware-software system exist. Using the rubric of artificial intelligence could complete the development of such an approach by creating the necessary predictive models, as I have advocated elsewhere [10].

6. Conclusion: health, well-being and HRV

It is an oversimplification to suggest that the adoption of the HRV software/hardware integrated system proposed would resolve the many serious issues that plague the current healthcare environment in the United States. Similarly, Topol of the Scripps Transformational Research Institute observes in his recent Deep Medicine [22] that the increasing use of artificial intelligence (AI) is not a panacea and can only contribute to improving the status quo. In particular, machine learning may make significant progress possible by using the relatively large number of data points generated by HRV and other psychophysiological parameters. It is a truism that both data quality and quantity are crucial in producing the most valuable predictive algorithmic equations. In addition to Topol's astute observations, other physicians such as Agus in the End of Illness [24] and Emanuel in Prescription for the Future [25] have also voiced more nuanced critiques of the healthcare venture in the US, identifying multiple domains of concern. The use of innovations such as machine learning and integrated HRV systems, however, may contribute to the achievement of a reformulated model of health, well-being and healthcare that is both more comprehensive and more tailored to the care of specific individuals. This approach was initially introduced as the human genome was being sequenced as personalized or precision medicine, the implication being that knowledge of the details of the individual's genetic makeup would make for highly specific treatment recommendations perhaps involving genetic engineering. A more multidimensional approach has emerged which addressed the many individual characteristics that each individual possesses; not just genetic, but biochemical, anatomical/physiological, cognitive/affective, social, cultural and spiritual. One of the major limitations of the majority of AI initiatives is the relative neglect of the affective domain of functioning that most distinctively characterizes humans. These multiple domains of individual variability require not only a highly personalized approach, but an integral view of the caring relationship that is participatory, predictive and preventive, as well. I have described this approach as the iP4 health model [10]. The emphasis on an *integral* perspective [26] highlights that each person is a complex adaptive system and that no aspect of their condition is independent from other details of their internal characteristics and external environmental conditions. The focus on *predictive* understanding acknowledges that a variety of risk and resilience factors exist and that any effort to maintain optimal health and well-being must note and plan to deal with such factors. Similarly, a *preventive* approach focuses on identification and early intervention to minimize or eliminate the onset and/or severity of disease states and promote health. Perhaps most important, we hope to partner with informed individuals and support their decisions that will maximize their *participation* in the care process, making it not only more efficient and effective, but also actually demonstrating care for the individual. It is only through a detailed integration of such important issues that a truly *personalized* health care is possible. Since we do not see this as a continuation of the existing approach to patient care, we prefer to designate the individual not as a patient, but a Pioneer.

The role which HRV may play in actualizing this model is expanding rapidly and, together with Drs. Steven Porges, Julian Thayer and Jay Ginsberg, I have edited a Research Topic that has jointly appeared in the journals Frontiers in Medicine and Frontiers in Public Health entitled "Heart Rate Variability, Health and Wellbeing: A Systems Perspective" [27]. This series of research papers and reviews summarizes current empirical findings and conceptual bases for applying our understanding of HRV to a wide variety of problems, diseases and issues. Thus, our efforts will not only monitor the various human nervous systems but help to assist and optimize them in their important task of shepherding each individual's health and well-being. Through both scientific and technological investigation using advanced AI tools such as machine learning, and appropriate provider education, a common interactional language must evolve that allows the evaluation of HRV and other relevant parameters to generate actionable feedback, whether decision support for physicians and other healthcare personnel or personal behavioral prompts or interventions for individuals. Though the current state of our understanding is relatively primitive, there is reason to be optimistic that such an evolution is possible.

Acknowledgements

The author acknowledges the valuable intellectual contributions of Drs. Wasyl Malyj, Steven Porges, Julian Thayer, Jack Ginsberg, Lee Hood, James Spira, Emma Seppala, Fritjof Capra, Steven Strogatz, Francisco Varela, Humberto Maturana and Sesha Robin Hanson-Drury (D.D.S., Ph.D. Cand).

Conflict of interest

The author declares no conflict of interest.

Author details

Robert L. Drury[1,2]

1 ReThink Health, Bainbridge Island, WA, USA

2 Institute for Discovery, University of Wisconsin, Madison, WI, USA

*Address all correspondence to: rl.drury@gmail.com

IntechOpen

© 2019 The Author(s). Licensee IntechOpen. This chapter is distributed under the terms of the Creative Commons Attribution License (http://creativecommons.org/licenses/by/3.0), which permits unrestricted use, distribution, and reproduction in any medium, provided the original work is properly cited. cc BY

References

[1] Porges S. The Polyvagal Theory. New York: Norton; 2011

[2] Thayer JF, Lane RD. Claude Bernard and the heart-brain connection: Further elaboration of a model of neurovisceral integration. Neuroscience and Biobehavioral Reviews. 2009;**33**(2):81-88

[3] Capra F, Luisi L. The Systems View of Life. Delhi: Cambridge University Press; 2014

[4] Kandel E et al. Neural Science. 5th ed. New York: McGraw; 2018

[5] Squire L et al. Fundamental Neuroscience. 4th ed. Burlington, MA: Academic Press; 2012

[6] Purves D et al. Neuroscience. 5th ed. Sunderland, MA: Sinauer; 2012

[7] Bainbridge FA. The relation between respiration and the pulse rate. The Journal of Physiology. 1920;**54**:192-202

[8] Cacioppo J et al. Handbook of Psychophysiology. 4th ed. New York: Cambridge University Press; 2016

[9] European Society of Cardiology and North American Society of Pacing and Electrophysiology. Heart rate variability: Standards of measurement, physiological interpretation and clinical use. Task force of the European Society of Cardiology and the north American Society of Pacing and Electrophysiology. Circulation. 1996;**93**:1043-1068

[10] Drury R. Unraveling health as a complex adaptive system: Data mining, cloud computing, machine learning, and biosensors as synergistic technologies. International Journal of Genomics and Data Mining. 2017;**J105**:1-4

[11] Sammito S, Bockelman E. Factors influencing heart rate variability. International Cardiovascular Forum Journal. 2016;**6**:18-22

[12] Thayer JF, Ahs F, Fredrikson M, Sollers JJ 3rd, Wager TD. A meta-analysis of heart rate variability and neuroimaging studies: Implications for heart rate variability as a marker of stress and health. Neuroscience and Biobehavioral Reviews. 2012;**36**(2):747-756

[13] Compos M. Heart Rate Variability: A New Way to Track Well-Being. Cambridge: Harvard Health Blog, Harvard Medical School; 2017

[14] Tarvainen MP et al. Kubios HRV-heart rate variability analysis software. Computer Methods and Programs in Biomedicine. 2014;**113**(1):210-220

[15] Haddad G. Medical Director. Seattle, WA: P4 Medical Institute. Personal Communication

[16] Liu NT et al. Utility of vital signs, heart-rate variability and complexity, and machine learning for identifying the need for life-saving interventions in trauma patients. Shock. 2014. pp. 10,1097

[17] King DR, Ogilvie MP, Pereira BM, Chang Y, Manning RJ, et al. Heart rate variability as a triage tool in patients with trauma during prehospital helicopter transport. The Journal of Trauma. 2009;**67**:436-440

[18] IBM Watson Analytics: A Smart Data Discovery Service Available on the Cloud, it Guides Data Exploration, Automates Predictive Analytics and Enables Effortless Dashboard and Infographic Creation. Available from: www.ibm.com [Accessed: 10 July 2019]

[19] Guidi G, Roberto M, Matteo M, Ernesto I. Case study: IBM Watson

analytics cloud platform as analytics-as-a-service system for heart failure early detection. Future Internet. 2016;**8**:32

[20] PhysioNet. PhysioNet is supported by the National Institute of General Medical Sciences (NIGMS) and the National Institute of Biomedical Imaging and Bioengineering (NIBIB) under NIH grant number 2R01GM104987-09. N.D. Available from: www.physionet.org [Accessed: 15 March 2017]

[21] Drury R. Wearable biosensor systems and resilience: A perfect storm in health care? Frontiers in Psychology. 2014;**5**:853

[22] Topol E. Deep Medicine. New York: Hatchett; 2019

[23] Drury R, Malyj W, Phares R. Resilience enhancement: A psychoeducational intervention using miniaturized electronic data collection and analysis. In: Presented at the American Medical Informatics Association Meeting ; San Francisco, CA; 2010

[24] Agus D. The End of Illness. New York: Free Press; 2012

[25] Emanuel E. Prescription for the Future. New York: Public Affairs; 2017

[26] Rakel D. Integrative Medicine. 4th ed. Philadelphia: Elsevier; 2019

[27] Drury R, Porges S, Thayer J, Ginsberg J, editors. Heart Rate Variability, Health and Wellbeing: A Systems Perspective. Special Research Topic in Frontiers in Medicine and Frontiers in Public Health, 2018

Chapter 2

Root Mean Square of the Successive Differences as Marker of the Parasympathetic System and Difference in the Outcome after ANS Stimulation

Giovanni Minarini

Abstract

The autonomic nervous system has a huge impact on the cardiac regulatory mechanism, and many markers exist for evaluating it. In this chapter we are going to focus on the RMSSD (Root mean square of successive differences), considered the most precise marker for the parasympathetic effector on the heart. Before is necessary to learn what the Heart Rate Variability is and how it works, which type of range of HRV exists and how we can measure it. Finally, there will be a presentation of how the RMSSD can be used in different field, and how and why the outcome can change and what does it mean.

Keywords: autonomic nervous system, sympathetic nervous system, parasympathetic nervous system, root mean square of successive differences, heart, cardiac mechanism, ANS influence

1. Introduction

Homeostasis can be defined as the result of the stability of physiological systems that maintain life; it applies strictly to a limited number of systems such as regulation of pH, concentration of different ions in the extracellular fluid, osmolality of extracellular fluid, glucose levels and arterial oxygen tension, which are truly essential for life and are therefore maintained within a narrow range for the current life history stage [1]. The homeostatic balance is considered as a change of state compatible with the actual environmental situations [2]. The temporary variations between the "set point" of the homeostatic control system during adaptation to internal (i.e. digestion) or external (i.e. climatic condition) perturbations are called *allostasis* [1]. Thus, allostasis is the reaching of physiological stability through a change of homeostatic state [1, 3].

The allostatic adaptations during environmental changes are temporary processes; if not turned off when not needed, if they occur too frequently or fail to occur at all, there may be development of systemic disease(s) such as cardiovascular disease, type II diabetes, obesity, etc. [4–6]. Allostatic and homeostatic control are ruled by the autonomic nervous system (ANS) integrated within the central

 IntechOpen

nervous system (CNS) [1]. A disorder of the ANS can affect the homeostatic and allostatic processes, leading to a risk of developing systemic disorder such as hypertension [7] baroreflex failure for blood pressure regulation [8], type II diabetes or affect the immune system and the inflammatory process [9]. Moreover, this has been demonstrated how the ANS is strictly correlated to the modulation of pain perceived by the subject.

One of the most common markers of the ANS is the rMSSD, square root of the mean squared differences of successive NN intervals. It is evaluated through the Heart Rate Variability (HRV) and the distance between the peaks of the R values in the echocardiogram. The rMSSD allowed the researcher to monitor the alteration in the parasympathetic activity with good precision. In this chapter will be explained how the rMSDD is related to parasympathetic nervous system, what could modulate the outcome and what a different result mean.

2. Heart rate variability

Heart rate variability consists of changes in the time intervals between consecutive heartbeats called interbeat intervals (IBIs) [10]. A healthy heart is not a metronome. The oscillations of a healthy heart are complex and constantly changing, which allow the cardiovascular system to rapidly adjust to sudden physical and psychological challenges to homeostasis.

The HRV is the fluctuation in time intervals between adjacent heartbeats, it indexes neurocardiac function and is generated by heart-brain interactions and dynamic non-linear autonomic nervous system (ANS) processes. HRV is an emergent property of interdependent regulatory systems which operate on different time scales to help us adapt to environmental and psychological challenges by stimulating and regulating some vascular component of the allostasis: HRV reflects regulation of autonomic balance, blood pressure (BP), gas exchange, gut, heart, and vascular tone, which refers to the diameter of the blood vessels that regulate BP, and possibly facial muscles [11].

Higher HRV is not always associated to better state of health of the subjects, numerous diseases affect the HRV and has the potential to increase this value. When cardiac conduction abnormalities cause an increase in the HRV, this is strongly linked to increased risk of mortality, particularly among the elderly (e.g. causing atrial fibrillation) [12].

Despite that has been demonstrated how optimal level of HRV are associated to health and self-regulatory capacity, adaptability and resilience [13, 14]. This is due to the vagal modulation of the HRV: heart rate (HR) estimated at any given time represents the net effect of the neural output of the parasympathetic (vagus) nerves, which slow HR, and the sympathetic nerves, which accelerate it. Opthof published a study on the normal range and the determinants of intrinsic heart rate in man, following the main research done before on the subject by Jose an Collins in 1970: he found that a denervated human heart, with no connections to the ANS, the intrinsic rate generated by the pacemaker, the Senoatrial Node (SA), is near to 100 beats per minute [15]. Whenever the rate decrease below this level, it means that a parasympathetic outflow is predominating in the balance between sympathetic and parasympathetic activity. This happens usually during normal daily activities, at rest and when we sleep. On the contrary, if the ratio raises over 100 beats the shift is toward the sympathetic system. The average 24 h HR in healthy people is approximately 73 bpm. Higher HRs are independent markers of mortality in a wide spectrum of conditions [16].

2.1 Sympathetic and parasympathetic pathways on the heart

Sinus node pacemaker cells activity is continuously under regulation by specific neural mechanism.

The SA node is targeted by the descending efferent sympathetic nerves via the intrinsic cardiac nervous system and the bulk of the myocardium. Norepinephrine and epinephrine release, which increases HR and strengthens the contractility of the atria and ventricles, is triggered by these motor neurons action potentials. Subsequent the onset of sympathetic stimulus, there is a delay of up to 5 seconds before the stimulation induces a progressive rise in HR, which reaches a stable level in 20–30 seconds if the stimulus is continuous [17]. A sympathetic stimulus, even if brief, can easily affect the HRV rhythm for 5–10 seconds. This is in contrast with the vagal stimulation, which is almost instantaneous, due to the acetylcholine degradation mechanism [17, 18], we will see that later on this chapter. What does that mean? That any sudden changes in the HR, up or down, or between the beat, is primarily mediated by the parasympathetic nervous system.

The vagus nerves innervate the intrinsic cardiac nervous system. Inside the intrinsic cardiac nervous systems are present some synapse between vagus nerve and motor neurons that directly project to the sinoatrial node and a portion of the surrounding tissue. They trigger acetylcholine release to slow HR. [19] However, more than 80% of the efferent preganglionic vagal neurons has connection to local circuitry neurons in the intrinsic cardiac nervous system where motor information is integrated with inputs from T.

The single efferent vagal stimulation on the SA node is very short, resulting in an immediate response that typically occurs within the cardiac circle in which it occurs, affecting only 1 or 2 heartbeats after its onset [17]. After cessation of vagal stimulation, HR rapidly increases to its previous level. An increase in heart rate can also be achieved by reduced vagal activity, or vagal withdrawal. Hence, any sudden increase or decrease HR, between 1 beat and the next, are primarily parasympathetically mediated [17, 18].

The medulla oblongata is the major structure integrating incoming afferent information from the heart, lungs and face with inputs from cortical and subcortical structures and is the source of the respiratory modulation of the activity patterns in sympathetic and parasympathetic outflow. The intrinsic cardiac nervous system integrates mechanosensitive and chemosensitive neuron inputs with efferent information from both the sympathetic and parasympathetic inputs from the brain. As a complete system, it affects HRV, vasoconstriction, venoconstriction, and cardiac contractility in order to regulate HR and BP [17].

2.2 Heart rate variability frequency band

The European Society of Cardiology and the North American Society of Pacing and Electrophysiology Task Force Report on HRV divided heart rhythm oscillations into 4 primary frequency bands: high-frequency (HF), low-frequency (LF), very-low-frequency (VLF), and ultra-low-frequency (ULF) [20]. Most HRV analysis is done in 5-min segments (of a 24 h recording), although other recording periods are often used. When other recording lengths are analyzed, the length of the recording should be reported since this has large effects on both HRV frequency and time domain values.

2.2.1 High-frequency band

The HF range is from 0.15 to 0.4 Hz, which correspond to a rhythm period between 2.5 and 7 seconds. This band is called the “respiratory band” because correspond to the HR variations related to the respiratory cycle, known also as "respiratory sinus arrhythmia". Is conventionally recorded over a minimum 1 min period. For infants and children, who breathe faster than adults, the resting range can be adjusted to 0.24–1.04 Hz [21]. A complex regulatory mechanism involving both central and reflex interaction is the main organizer of this system. During inhalation there is an acceleration in the heart rate due to an inhibition of the vagal/ outflow from the cardio-respiratory center. On the opposite, while exhaling, the vagal outflow is restored to a normal level resulting in slowing the HR.

The HF modulation has also psychological involvement: reduced vagally mediated HRV has been related to a reduced self-regulatory capacity and cognitive functions that involve the executive centers of the prefrontal cortex. This is consistent with the finding that lower HF power is associated with stress, panic, anxiety, or worry. It has to be noted how this reaction is due a reduction of the parasympathetic activity, and not to an increase in sympathetic ones. This has been shown by numerous studies, where a total vagal blockade obtained pharmacologically eliminates the HF oscillations and reduces power in the LF range, resulting in a strong reduction of the HRV, including LF and VLF bands. Thus, they concluded that HRV is a resultant of the parasympathetic mechanism.

High-frequency power is highly correlated with the pNN50 and RMSSD time-domain measures [22]. HF band power may increase at night and decrease during the day [10].

2.2.2 Low-frequency band

LF range is between 0.04 and 0.15 Hz, which equates to rhythms or modulations periods between 7 and 25 seconds. Is typically recorded over a minimum 2 min period [23]. This range of action was called the "baroreceptor range" or "mid-frequency band", due to its strong correlation with baroreceptor activity at rest [10, 24]. Baroreceptors are stretch-sensitive mechanoreceptors located in vena cavae, carotid sinuses, aortic arch and heart chambers. The ones found in the carotid are the most sensitive. Baroreflex is transported to the brain by the vagus nerve and represent the beat-to-beat change in HR per unit of change in systolic blood pressure [25]. A decreased baroreflex is related to aging and weakened regulatory capacity [26].

There is a different influence of the sympathetic and parasympathetic system inside this band, due to the rhythms: above 0.1 Hz the SNS seems to be lesser influent, whilst the parasympathetic affect heart rhythms down to 0.05 Hz [27, 28]. There rhythms are obtained during slow respiration rates, where a vagal activity easily generates oscillations in the heart rhythms crossing into the LF band [29, 30]. Therefore, when the respiratory rates are below 8.5 per minutes, or 1 in 7 seconds, or when a subject take a deep breath there is a vagal mediation influence.

Despite has been generally accepted the LF band has a marker for the sympathetic activity, and the LF/HF ratio is used to assess the balance between SNS and PNS, is still not totally clear if in resting condition the Low Frequency band reflect the baroreflex activity instead of the cardiac sympathetic innervation [31–33].

2.2.3 Very-low-frequency band

The VLF is the power in the range between 0.0033 and 0.04 Hz, which equates to rhythms or modulations with periods that occur between 25 and 300 seconds. Although all 24 h clinical measures of HRV reflecting low HRV are linked with increased risk of adverse outcomes, the VLF band has stronger associations with all-cause mortality than the LF and HF bands [34–37]. Low VLF power has been shown to be associated with arrhythmic death [38] and posttraumatic stress disorder (PTSD) [39]. Moreover, low power expression in this band has been associated with high inflammation [40, 41] and has been correlated with low levels of testosterone. In contrast, other biochemical markers, such as those mediated by the hypothalamic–pituitary–adrenal (HPA) *Axis axis* (e.g., cortisol), did not [42]. Longer time periods using 24 h HRV recordings should be obtained to provide comprehensive assessment of VLF and ULF fluctuations [22].

Historically, is still not well defined the physiological explanation and mechanisms involved in the generation of the VLF component, compared to the LF and HF components. Despite the accuracy and the most predictive outcomes, this area has been largely ignored even. Long-term regulatory mechanisms and ANS activity related to thermoregulation, the renin-angiotensin system, and other hormonal factors appear to contribute to this band [43, 44].

Very-low-frequency power is strongly correlated with the SDNNI time-domain measure, which averages 5 min standard deviations for all NN intervals over a 24 h period. There is uncertainty regarding the physiological mechanisms responsible for activity within this band [14]. The heart's intrinsic nervous system appears to contribute to the VLF rhythm and the SNS influences the amplitude and frequency of its oscillations [20].

Based on work by Armor [45] and Kember et al. [32, 46], the VLF rhythm appears to be generated by the stimulation of afferent sensory neurons in the heart. This, in turn, activates various levels of the feedback and feed-forward loops in the heart's intrinsic cardiac nervous system, as well as between the heart, the extrinsic cardiac ganglia, and spinal column. This experimental evidence suggests that the heart intrinsically generates the VLF rhythm and efferent SNS activity due to physical activity and stress responses modulates its amplitude and frequency.

2.2.4 Ultra-low-frequency band

The ultra-low-frequency band (ULF) falls below 0.0033 Hz (333 seconds or 5.6 minutes). Oscillations or events in the heart rhythm with a period of 5 minutes or greater are reflected in this band and it can only be assessed with 24 h and longer recordings [22]. The circadian oscillation in HR is the primary source of the ULF power, although other very slow-acting regulatory processes, such as core body temperature regulation, metabolism, and the renin-angiotensin system likely add to the power in this band [20]. The Task Force Report on HRV suggests that 24 h recordings should be divided into 5-min segments and that HRV analysis should be performed on the individual segments prior to the calculation of mean values. This effectively filters out any oscillations with periods longer than 5 minutes. However, when spectral analysis is applied to entire 24 h records, several lower frequency rhythms are easily detected in healthy individuals [23].

There is disagreement about the contribution by the PNS and SNS to this band. Different psychiatric disorders show distinct circadian patterns in 24 h HRs, particularly during sleep [25, 47].

2.3 Heart rate variability measurement

Three types of measurement exist for the HRV, time-domain index, frequency-domain index and non-linear measurements. Time-domain indices quantify the amount of HRV observed during monitoring periods that may range from ~2 min to 24 h. Frequency-domain values calculate the absolute or relative amount of signal energy within component bands. *Non-linear measurements* allow us to quantify the unpredictability of a time series.

2.3.1 Time domain measurements of heart rate variability

Time domain measures are the simplest to calculate. Time domain measures do not provide a means to adequately quantify autonomic dynamics or determine the rhythmic or oscillatory activity generated by the different physiological control systems. However, since they are always calculated the same way, data collected by different researchers are comparable but only if the recordings are exactly the same length of time and the data are collected under the same conditions. Time domain indices quantify the amount of variance in the inter-beat-intervals (IBI) using statistical measures. The three most important and commonly reported time domain measures are the standard deviation of normal-to-normal (SDNN), the SDNN index, and the root mean square of successive differences (RMSSD) are the most commonly reported metrics.

2.3.2 The standard deviation of the normal-to-normal

The SDNN is the standard deviation of the normal-to-normal (NN) sinus-initiated IBIs measured in milliseconds. This measure reflects the ebb and flow of all the factors that contribute to HRV. In 24 h recordings, the SDNN is highly correlated with ULF and total power [48]. In short-term resting recordings, the primary source of the variation is parasympathetically mediated, especially with slow, deep breathing protocols. However, in ambulatory and longer term recordings the SDNN values are highly correlated with lower frequency rhythms [23]. Thus, low age-adjusted values predict morbidity and mortality. For example, patients with moderate SDNN values (50–100 milliseconds) have a 400% lower risk of mortality than those with low values (0–50 milliseconds) in 24 h recordings [49, 50].

2.3.3 Standard deviation of the normal-to-normal index

The SDNN index is the mean of the standard deviations of all the NN intervals for each 5 min segment. Therefore, this measurement only estimates variability due to the factors affecting HRV within a 5 min period. In 24 h HRV recordings, it is calculated by first dividing the 24 h record into 288 five-minute segments and then calculating the standard deviation of all NN intervals contained within each segment. The SDNN index is the average of these 288 values [20]. The SDNN index is believed to primarily measure autonomic influence on HRV. This measure tends to correlate with VLF power over a 24 h period [23].

2.3.4 The root mean square of successive differences

The RMSSD is the root mean square of successive differences between normal heartbeats. This value is obtained by first calculating each successive time difference between heartbeats in milliseconds. Each of the values is then squared and the result is averaged before the square root of the total is obtained. The RMSSD reflects the beat-to-beat variance in HR and is the primary time domain measure used to estimate the vagally mediated changes reflected in HRV [20]. The RMSSD is correlated with HF power and therefore also reflects self-regulatory capacity as discussed earlier [23].

3. Root mean square of successive differences as PNS marker

As aforementioned, the RMSSD reflects the vagally mediated changes in the relation that occur between two peaks in the R value of an echocardiogram, thus give to the researcher an overview of the PNS activity.

The parasympathetic modification evaluation is one of the most used parameters and find its utility in different research [51, 52]. Accordingly, to Zygmunt and Stanczyk [53], the rMSSD *"describes short-term variations, and thus reflects parasympathetic activities"*. Although the HRV and rMSSD does not reflect perfectly the tonic activity of parasympathetic and sympathetic, but rather the resultant on the effector, that are sinus cells node receptors; the vagal activity is predominant compared to sympathetic one. Influence of parasympathetic is fast and transient, due to acetylcholine degradation by acetylesterase. These physiological redundancies cause parasympathetic activities to be visible in the cycle that follows the stimulus, whilst the sympathetic stimulation develop more slowly, and their effects are visible only after 2–3 s, causing slower oscillation with higher amplitude [54]. One concern regarding this issue is that HRV studies are quite sensitive to a number of factors as eloquently pointed out by Piché and Descarreaux [55], which can make data interpretation challenging.

Due to the ambiguity in physiological meaning in low frequency (LF) variations during short recording periods [20] the Time Domain Indices (rMSSD) has revealed itself more reliable than frequency domain [56], and considered as PNS modulation indices [20]. R-R intervals is a time domain measure of HRV calculated by the equation of Kim et al. [57]. According to Hayward et al. [58], the rMSSD time domain measurement has high sensitivity to identify ANS modification in temporal window of 1–2 minutes, concordant to Thong et al. [59] who found how rMSSD is a valuable measurement for ultra-short-term records (1–5 minutes) due to its ability to be improved by combining disjoint records; e.g. combining 6 rMSSD records of 10 seconds each to obtain the equivalent of a 60 seconds length rMMSD registration. Moreover, Esco and Flat [60] showed and almost perfect relationship between ultra-short-term and criterion measures (5 minutes) by recording rMSSD in 23 athletes pre- and post-exercises.

3.1 RMSSD modulation with physical activity

RMSSD is often used for professional athletes in order to monitor cardiac activity and modulation of the HRV subsequentially to physical performance. Acute decreases in HRV have been reported to occur following intense endurance training [61], resistance training [62], and competition [63]. Therefore, low HRV is commonly thought to provide a reflection of acute fatigue from training or competing.

But despite what has been accepted for the last years, recent discovery shows how not always an increase in this value is a positive result.

In the context of monitoring fatigue or training status in athletes, a common belief is that high HRV is good and low HRV is bad. Or, in terms of observing the overall trend, increasing HRV trends are good, indicative of positive adaptation or increases in fitness. Decreasing trends are bad, indicative of fatigue accumulation or "overtraining" and performance decrements.

Unfortunately, an increasing HRV trend throughout training is not always a good thing and thus should not always be interpreted as such. In fact, several studies have reported increasing HRV trends in overtrained athletes predominately involved in endurance sports. For example, Le Meur et al. [64] showed decreased maximal incremental exercise performance and increased weekly HRV mean values in elite endurance athletes following a 3 week overload period, compared to a control group who saw no changes. Following a taper, performance supercompensation was observed along with a return of HRV toward baseline.

He most common response to overload training is a progressive decrease in HRV. This is your typical alarm response to a stressor, where the sympathetic arm of the autonomic nervous system is activated. In this situation, resting HR is elevated and HRV decreases. With insufficient recovery time, HRV may not fully recover to baseline before the next training stimulus and thus will result in a downward trend when this cycle is perpetuated. An intense day of training can result in suppressed HRV for up to 72 hours post-exercise [65]. With the higher training frequencies and training volumes often associated with overload periods, it makes sense that HRV will show a decreasing trend. Typically, HRV will respond first with a decreasing trend and performance decrements will follow if the overload period is sustained.

A study by Pichot et al. [66] provides a good example of a decreasing HRV trend in response to overload training. They showed that middle distance runners saw a progressive downward HRV trend (up to −43%) during a 3 week overload period. In week 4, training loads were reduced and HRV recovered and exceeded baseline values.

These aspects demonstrate a new aspect of the RMSSD modulation due to a physical stimulus, and the complexity of the cardiac regulation mechanism.

3.2 RMSSD application inside the psychological field

Altered cardiac autonomic functions in form of reduced Heart Rate Variability (HRV) have been found to be associated with increased cardiovascular morbidity and mortality in depressive patients.

Most studies have now identified depression as a strong and independent risk factor for cardiovascular disease even in physically healthy individuals [67] and also for adverse cardiovascular outcomes such as mortality [68]. Although the underlying pathophysiological mechanism is yet to be elucidated, autonomic imbalance has been projected as one of the underlying mechanisms [69]. Heart Rate Variability (HRV) is a useful non-invasive measure for assessing cardiac autonomic modulations.

Reduced HRV has been reported in several studies done in depressed patients both with and without cardiovascular diseases compared to non-depressed subjects [70, 71]. Although negative studies have been reported as well, which were unable to prove an association [72]. Most of the researches carried out to observe the association between HRV and depression have been done in individuals who were already either having Cardiovascular Disease (CVD) besides depression or were on medications [73].

A meta-analysis done by Kemp et al., in depression patients without cardiovascular diseases also reported the association between reduced HRV and depression and was found to be more in severely depressed individuals [74].

In addition, Agelink et al., showed the inverse correlation of parasympathetic HRV values with the severity of depression [75]. In a recent study Wang et al., also observed higher LF, LF:HF Ratio and lower SDNN, RMSSD and HF values in depression group compared to control group [52].

4. Conclusion

The rMSSD can be a very useful tool for many relevant findings: from the parasympathetic activation to a marker for cardiac dysfunction. Despite the findings and the large are of application, it is still an unknown area.

The correlation between the psychological issues and the rMSSD value add a deeper meaning on how the body is strictly correlated to the mind, and the interrelation between the thought and its physical response. Many research has been done regarding the heart and its physiology and mechanism, but only now we are starting to really understand how this tissue really works, and for better comprehend how this organ is connected to the body and how it respond to the numerous different stimuli throughout the day, is necessary to delineate a clear structure of evaluation capable of considering also these new aspect, like the psychological impact and the psychological response.

Author details

Giovanni Minarini
Studio Alphaomega, Modena, Italy

*Address all correspondence to: giovanni.minarini@gmail.com

IntechOpen

© 2020 The Author(s). Licensee IntechOpen. This chapter is distributed under the terms of the Creative Commons Attribution License (http://creativecommons.org/licenses/by/3.0), which permits unrestricted use, distribution, and reproduction in any medium, provided the original work is properly cited. cc BY

References

[1] McEwen BS, Wingfield JC. The concepts of allostasis in biology and biomedicine. Hormones and Behavior. 2003;**43**:2-15

[2] Janig W. Homeostasis and Allostasys. In: The Integrative Action of the Autonomic Nervous System. Vol. 2. Cambridge: Cambridge University Press; 2008. 1p

[3] Schulkin J. Allostasis: A neural behavioral perspective. Hormones and Behavior. 2003;**43**:21-27

[4] Bjorntop P. Behavior and metabolic disease. International Journal of Behavioral Medicine. 1997;**2**:285-302

[5] Folkow B, Schmidt T, Uvnas-Moberg K. Stress, health and the social environment. Acta Physiologica Scandinavica. 1997;**161**(640):1-179

[6] McEwen BS. Neurobiology of interpreting and responding to stressful events: Paradigmatic role of the hippocampus. In: Handbook of Physiology. Section 7: The Endocrine System, Vol. IV, Coping with the Environment: Neural and Neuroendocrine Mechanisms. Oxford New York: Oxford University Press; 2001. pp. 155-178

[7] Grassi G, Serravalle G. Sympatho-vagal imbalance in hypertension. In: Robertson D, Biaggioni I, Burnstock G, Low PA, Paton JFR, editors. Primer on the Autonomic Nervous System. Amsterdam: Elsevier; 2012

[8] Jordan J. Baroreflex failure. In: Robertson D, Biaggioni I, Burnstock G, Low PA, Paton JFR, editors. Primer on the Autonomic Nervous System. Croydon, UK: Academic Press; 2012. pp. 349-353

[9] Marvar PJ, Harrison DG. Inflammation, immunity and the autonomic nervous system. In: Robertson D, Biaggioni I, Burnstock G, Low PA, Paton JFR, editors. Primer on the Autonomic Nervous System. Amsterdam: Elsevier; 2012

[10] McCraty R, Shaffer F. Heart rate variability: New perspectives on physiological mechanisms, assessment of self-regulatory capacity, and health risk. Global Advances in Health and Medicine. 2015;**4**(1):46-61

[11] Gevirtz RN, Lehrer PM, Schwartz MS. Cardiorespiratory biofeedback. In: Schwartz MS, Andrasik F, editors. Biofeedback: A Practitioner's Guide. 4th ed. New York: The Guilford Press; 2016. pp. 196-213

[12] Stein PK, Domitrovich PP, Hui N, Rautaharju P, Gottdiener J. Sometimes higher heart rate variability is not better heart rate variability: Results of graphical and nonlinear analyses. Journal of Cardiovascular Electrophysiology. 2005;**16**(9):954-959

[13] McCraty R, Atkinson M, Tomasino D, Bradley RT. The Coherent Heart: Heart-Brain Interactions, Psychophysiological Coherence, and the Emergence of System-Wide Order. Boulder Creek, CA: Institute of Heartmath; 2009

[14] Segerstrom SC, Nes LS. Heart rate variability reflects self-regulatory strength, effort, and fatigue. Psychological Science. 2007;**18**(3):275-281

[15] Opthof T. The normal range and determinants of the intrinsic heart rate in man. Cardiovascular Research. 2000;**45**(1):173-176

[16] Palatini P. Elevated heart rate as a predictor of increased cardiovascular morbidity. Journal of Hypertension. Supplement. 1999;**17**(3):S3-S10

[17] Hainsworth R. The control and physiological importance on heart rate. In: Malik M, Camm AJ, editors. Heart Rate Variability. Armonk, NY: Futura Publishing Company, Inc; 1995

[18] Lacey BC, Lacey JI. Two-way communication between the heart and the brain. Significance of time within the cardiac cycle. The American Psychologist. 1978;**33**(2):99-113

[19] Armour JA. Anatomy and function of the intrathoracic neurons regulating the mammalian heart. In: Zucker IH, Gilmore JP, editors. Reflex Control of the Circulation. Boca Raton, FL: CRC Press; 1991. pp. 1-37

[20] Task Force of the European Society of Cardiology, North American Society of Pacing and Electrophysiology. Heart rate variability standards of measurement, physiological interpretation, and clinical use. Circulation. 1996;**93**(5):1043-1065

[21] Quintana DS, Elstad M, Kaufmann T, Brandt CL, Haatveit B, Haram M, et al. Resting-state high-frequency heart rate variability is related to respiratory frequency in individuals with severe mental illness but not healthy controls. Scientific Reports. 2016;**6**:37212. DOI: 10.1038/srep37212

[22] Kleiger RE, Stein PK, Bigger JT. Heart rate variability: Measurement and clinical utility. Annals of Noninvasive Electrocardiology. 2005;**10**:88-101. DOI: 10.1111/j.1542-474X.2005.10101.x

[23] Shaffer F, McCraty R, Zerr CL. A healthy heart is not a metronome: An integrative review of the heart's anatomy and heart rate variability. Frontiers in Psychology. 2014;**5**:1040. DOI: 10.3389/fpsyg.2014.01040ù

[24] Malliani A. Association of heart rate variability components with physiological regulatory mechanisms. In: Malik M, Camm AJ, editors. Heart Rate Variability. Armonk NY: Futura Publishing Company, Inc; 1995. pp. 173-188

[25] Bowers EJ, Murray A. Effects on baroreflex sensitivity measurements when different protocols are used to induce regular changes in beat-to-beat intervals and systolic pressure. Physiological Measurement. 2004;**25**(2):523-538

[26] Bristow JD, Gribbin B, Honour AJ, Pickering TG, Sleight P. Diminished baroreflex sensitivity in high blood pressure and ageing man. The Journal of Physiology. 1969;**202**(1):45-46

[27] de Boer RW, Karemaker JM, Strackee J. Hemodynamic fluctuations and baroreflex sensitivity in humans: A beat-to-beat model. The American Journal of Physiology. 1987;**253**(3 Pt 2):H680-H689

[28] Baselli G, Cerutti S, Badilini F, Biancardi L, Porta A, Pagani M, et al. Model for the assessment of heart period and arterial pressure variability interactions and of respiration influences. Medical & Biological Engineering & Computing. 1994;**32**(2):143-152

[29] Ahmed AK, Harness JB, Mearns AJ. Respiratory control of heart rate. European Journal of Applied Physiology. 1982;**50**:95-104

[30] Tiller WA, McCraty R, Atkinson M. Cardiac coherence: A new, noninvasive measure of autonomic nervous system order. Alternative Therapies in Health and Medicine. 1996;**2**(1):52-65

[31] Eckberg DL. Human sinus arrhythmia as an index of vagal cardiac outflow. Journal of Applied Physiology: Respiratory, Environmental and Exercise Physiology. 1983;**54**(4):961-966

[32] Kember GC, Fenton GA, Armour JA, Kalyaniwalla N. Competition model for aperiodic stochastic resonance in a Fitzhugh-Nagumo model of cardiac sensory neurons. Physical Review. E, Statistical, Nonlinear, and Soft Matter Physics. 2001;**63**(4 Pt 1):041911

[33] MacKinnon S, Gevirtz R, McCraty R, Brown M. Utilizing heartbeat evoked potentials to identify cardiac regulation of vagal afferents during emotion and resonant breathing. Applied Psychophysiology and Biofeedback. 2013;**38**(4):241-255

[34] Tsuji H, Venditti FJ, Manders ES, Evans JC, Larson MG, Feldman CL, et al. Reduced heart rate variability and mortality risk in an elderly cohort. The Framingham heart study. Circulation. 1994;**90**(2):878-883

[35] Schmidt H, Müller-Werdan U, Hoffmann T, Francis DP, Piepoli MF, Rauchhaus M, et al. Autonomic dysfunction predicts mortality in patients with multiple organ dysfunction syndrome of different age groups. Critical Care Medicine. 2005;**33**(9):1994-2002

[36] Hadase M, Azuma A, Zen K, et al. Very low frequency power of heart rate variability is a powerful predictor of clinical prognosis in patients with congestive heart failure. Circulation Journal. 2004;**68**(4):343-347

[37] Tsuji H, Larson MG, Venditti FJ Jr, Manders ES, Evans JC, Feldman CL, et al. Impact of reduced heart rate variability on risk for cardiac events. The Framingham heart study. Circulation. 1996;**94**(11):2850-2855

[38] Bigger JT Jr, Fleiss JL, Steinman RC, Rolnitzky LM, Kleiger RE, Rottman JN. Frequency domain measures of heart period variability and mortality after myocardial infarction. Circulation. 1992;**85**(1):164-171

[39] Shah AJ, Lampert R, Goldberg J, Veledar E, Bremner JD, Vaccarino V. Posttraumatic stress disorder and impaired autonomic modulation in male twins. Biological Psychiatry. 2013;**73**(11):1103-1110

[40] Lampert R, Bremner JD, Su S, et al. Decreased heart rate variability is associated with higher levels of inflammation in middle-aged men. American Heart Journal. 2008;**156**(4):759.e1-759.e7

[41] Carney RM, Freedland KE, Stein PK, et al. Heart rate variability and markers of inflammation and coagulation in depressed patients with coronary heart disease. Journal of Psychosomatic Research. 2007;**62**(4):463-467

[42] Theorell T, Liljeholm-Johansson Y, Björk H, Ericson M. Saliva testosterone and heart rate variability in the professional symphony orchestra after "public faintings" of an orchestra member. Psychoneuroendocrinology. 2007;**32**(6):660-668

[43] Akselrod S, Gordon D, Ubel FA, Shannon DC, Berger AC, Cohen RJ. Power spectrum analysis of heart rate fluctuation: A quantitative probe of beat-to-beat cardiovascular control. Science. 1981;**213**(4504):220-222

[44] Cerutti S, Bianchi AM, Mainardi LT. Spectral analysis of the heart rate variability signal. In: Malik M, Camm AJ, editors. Heart Rate Variability. Armonk NY: Futura Publishing Company, Inc; 1995. pp. 63-74

[45] Armour JA. Neurocardiology: Anatomical and Functional Principles. Boulder Creek, CA: Institute of HeartMath; 2003

[46] Kember GC, Fenton GA, Collier K, Armour JA. Aperiodic stochastic resonance in a hysteretic

population of cardiac neurons. Physical Review E. 2000;**61**:1816-1824. DOI: 10.1103/PhysRevE.61.1816

[47] Armour JA. Peripheral autonomic neuronal interactions in cardiac regulation: Neurocardiology. In: Armour JA, Ardell JL, editors. New York: Oxford University Press; 1994. pp. 219-244

[48] Umetani K, Singer DH, McCraty R, Atkinson M. Twenty-four hour time domain heart rate variability and heart rate: Relations to age and gender over nine decades. Journal of the American College of Cardiology. 1998;**31**(3):593-601

[49] Stein PK, Domitrovich PP, Huikuri HV, Kleiger RE. Cast investigators. Traditional and nonlinear heart rate variability are each independently associated with mortality after myocardial infarction. Journal of Cardiovascular Electrophysiology. 2005;**16**(1):13-20

[50] Kleiger RE, Miller JP, Bigger JT Jr, Moss AJ. Decreased heart rate variability and its association with increased mortality after acute myocardial infarction. The American Journal of Cardiology. 1987;**59**(4):256-262

[51] Koenig J, Jarczok MN, Ellis RJ, Warth M, Hillecke TK. Lowered parasympathetic activity in apparently healthy subjects with self-reported symptoms of pain: Preliminary results from a pilot study. Pain Practice. 2015;**15**:314-318

[52] Wang Y, Zhao X, O'Neil A, Turner A, LiuEmail X. Altered cardiac autonomic nervous function in depression. BMC Psychiatry. 2013;**13**:187

[53] Zygmunt A, Stanczyk J. Methods of evaluation of autonomic nervous system function. Archives of Medical Science. 2010;**6**(1):11-18

[54] Pinna GD, Maestri R, Torunski A. Heart rate variability measures: A fresh look at reliability. Clinical Science. 2007;**113**:131-140

[55] Piché M, Descarreaux M. Heart rate variability modulation after manipulation in pain-free patients vs patients in pain? The importance of controlling for respiration rate changes. Journal of Manipulative and Physiological Therapeutics. 2010;**33**:554-555

[56] Al Haddad H, Laursen PB, Chollet D, Ahmaidi S, Buchheit M. Reliability of resting and postexercise heart rate measures. International Journal of Sports Medicine. 2011;**32**:598-605

[57] Kim KK, Lim YG, Kim JS, Park KS. Effect of missing RR-interval data on heart rate variability analysis in the time domain. Physiological Measurement. 2007;**28**(12):1458-1494

[58] Hayward C, Patel H, Manohar S, Waton J, Middleton L, Wardle A, et al. Heart rate variability, assessed in one minute windows, provides insight into the time course of changes in autonomic nervous system activity. Journal of the American College of Cardiology. 2015;**65**(10)

[59] Thong T, Li K, McNames J, Aboy M, Goldstein B. Accuracy of ultra-short heart rate variability measures. In: Proceedings of the 25th Annual International Conference of the IEEE Engineering in Medicine and Biology Society. Vol. 3. 2003. pp. 2424-2427

[60] Esco MR, Flatt AA. Ultra-short-term heart rate variability indexes at rest and post-exercise in athletes: Evaluating the agreement with accepted recommendations. Journal of Sports Science and Medicine. 2014;**13**:535-541

[61] Plews DJ, Laursen PB, Kilding AE, Buchheit M. Heart rate variability and

training intensity distribution in elite rowers. International Journal of Sports Physiology and Performance. 2014

[62] Chen JL, Yeh DP, Lee JP, Chen CY, Huang CY, Lee SD, et al. Parasympathetic nervous activity mirrors recovery status in weightlifting performance after training. The Journal of Strength & Conditioning Research. 2011;**25**(6):1546-1552

[63] Edmonds RC, Sinclair WH, Leicht AS. Effect of a training week on heart rate variability in elite youth rugby league players. International Journal of Sports Medicine. Dec 2013;**34**(12):1087-1092

[64] Le Meur Y, Pichon A, Schaal K, Schmitt L, Louis J, Gueneron J, et al. Evidence of parasympathetic hyperactivity in functionally overreached athletes. Medicine and Science in Sports and Exercise. 2013;**45**(11):2061-2071

[65] Stanley J, Peake JM, Buchheit M. Cardiac parasympathetic reactivation following exercise: Implications for training prescription. Sports Medicine. 2013;**43**(12):1259-1277

[66] Pichot V, Roche F, Gaspoz JM, Enjolras F, Antoniadis A, Minini P, et al. Relation between heart rate variability and training load in middle-distance runners. Medicine and Science in Sports and Exercise. 2000;**32**(10):1729-1736

[67] Gan Y, Gong Y, Tong X, Sun H, Cong Y, Dong X, et al. Depression and the risk of coronary heart disease: A meta-analysis of prospective cohort studies. BMC Psychiatry. 2014;**14**:371

[68] Freedland KE, Carney RM. Depression as a risk factor for adverse outcomes in coronary heart disease. BMC Medicine. 2013;**11**:131

[69] Huffman JC, Celano CM, Beach SR, Motiwala SR, Januzzi JL. Depression and cardiac disease: Epidemiology, mechanisms and diagnosis. Cardiovascular Psychiatry and Neurology. 2013;**2013**:1-14

[70] Glassman AH, Bigger JT, Gaffney M, Vanzyl LT. Heart rate variability in acute coronary syndrome patients with major depression: Influence of sertraline and mood improvement. Archives of General Psychiatry. 2007;**64**:1025-1031

[71] Kemp AH, Quintana DS, Felmingham KL, Matthews S, Jelinek HF. Depression, comorbid anxiety disorders and heart rate variability in physically healthy, unmedicatedpatients: Implications for cardiovascular risk. PLoS One. 2012;7:e30777

[72] Gehi A, Mangano D, Pipkin S, Browner WS, Whooley MA. Depression and heart rate variability in patients with stable coronary heart disease: Findings from the heart and soul study. Archives of General Psychiatry. 2005;**62**:661-666

[73] Carney RM, Freedland KE. Depression and heart rate variability in patients with coronary heart disease. Cleveland Clinic Journal of Medicine. 2009;**76**:13-17

[74] Kemp AH, Quintana DS, Gray MA, Felmingham KL, Brown K, Gatt JM. Impact of depression and antidepressant treatment on heart rate variability: A review and meta-analysis. Biological Psychiatry. 2010;**67**:1067-1074

[75] Agelink MW, Majewski T, Wurthmann C, Postert T, Linka T, Rotterdam S, et al. Autonomic neurocardiac function in patients with major depression and effects of antidepressive treatment with nefazodone. Journal of Affective Disorders. 2001;**62**:187-198

Chapter 3

Heart Rate Variability Recording System Using Photoplethysmography Sensor

Noor Aimie-Salleh, Nurul Aliaa Abdul Ghani, Nurhafiezah Hasanudin and Siti Nur Shakiroh Shafie

Abstract

Heart rate variability (HRV) is a physiological measurement that can help to monitor and diagnose chronic diseases such as cardiovascular disease, depression, and psychological stress. HRV measurement is commonly extracted from the electrocardiography (ECG). However, ECG has bulky wires where it needs at least three surface electrodes to be placed on the skin. This may cause distraction during the recording and need longer time to setup. Therefore, photoplethysmography (PPG), a simple optical technique, was suggested to obtain heart rate. This study proposes to investigate the effectiveness of PPG recording and derivation of HRV for feature analysis. The PPG signal was preprocessed to remove all the noise and to extract the HRV. HRV features were collected using time-domain analysis (TA), frequency-domain analysis (FA) and nonlinear time-frequency analysis (TFA). Five out of 22 HRV features, which are HR, RMSSD, LF/HF, LFnu, and HFnu, showed high correlation (rho > 0.6 and prho < 0.05) in comparison to standard 5-min excerpt while producing significant difference (p-value < 0.05) during the stressing condition across all interval HRV excerpts. This simple yet accurate PPG recording system perhaps might useful to assess the HRV signal in a short time, and further can be used for the ANS assessment.

Keywords: HRV, PPG, stress, autonomic function, ECG

1. Introduction

Human body is interacting between each other where it consists of many different interacting systems. Any changes in human body will generate response to all parts of the body include the autonomic nervous system (ANS) [1]. ANS controls the system that regulates bodily functions such as the digestion, respiratory rate, heart rate, pupillary response, urination, and sexual arousal. Any changes in ANS can be detected by heart rate variability (HRV) since HRV and ANS is directly related.

Heart rate can be defined as the number of heart beats per minute while heart rate variability (HRV) is the fluctuation in the time intervals between adjacent heartbeats. HRV refers to the time series of the interval variation between consecutive heart beats and it can be analyzed in time, frequency and nonlinear domains [2]. The fluctuations in HRV value reflects neurocardiac function of the body as it is generated through heart-brain connection and autonomic nervous system (ANS) dynamics [3, 4].

HRV is a common measurement that can be extracted from the physiological measurement and helps to monitor the psychological stress [5]. It is because, HRV has direct connection with the autonomic nervous system (ANS) where any changes that occurred in human body can be directly detected by the HRV. The common methods to get the HRV are by using the ECG. However, there are several difficulties to record the ECG signal. First, it requires at least three surface electrodes to be placed on the skin to get single lead channel [6]. This clearly shows bulky of wires are needed for the recording and might cause distraction and uncomfortable feeling to the patient. Furthermore, it requires several times to set up the ECG before start the recording.

In deriving the HRV signal, appropriate QRS algorithms need to be applied to detect the peaks and its R wave, to obtain the interval of RR, and to find acceptable interpolation and resampling to produce a consistently sampled tachogram. By using the ECG signal, the resultant HRV could have several errors in the HRV signal due to drift, electromagnetic and biological disturbance, and the complicated morphology of the ECG signal [6].

Therefore, a simple recording system in deriving the HRV signal is needed. PPG which is an electro-optical technique that detect the changes of blood volume in the microvascular bed of the tissue is believed able to overcome the problem that faced by ECG signal and has been suggested as an alternative method to derive the HRV signal [7].

The PPG sensor's system is equipped with a light source and a detector, it also developed with red and infrared (IR) light-emitting diodes (LEDs) that commonly used as the light source. The light intensity of the PPG sensor monitor has been changed via the reflection from or transmission through the tissue. **Figure 1** shows the signal from ECG and PPG signal. Derivation HRV signal from ECG is calculated from R-R interval, while the calculation of HRV signal from PPG signal is used inter-beat interval (IBI) or pulse interval (PPI) [8].

The light traveling through biological tissue passes many materials, including pigments in the skin, bone, and arterial and venous blood. The changes of blood flow mainly occur in the arteries and arterioles (but not in the veins). For example, during the systolic phase of the cardiac cycle, the arteries contain more blood volume than the diastolic phase. PPG sensors optically detect changes in the blood flow volume, for instance, changes in the detected light intensity in the microvascular bed of tissue through the reflection from or transmission through the tissue [9].

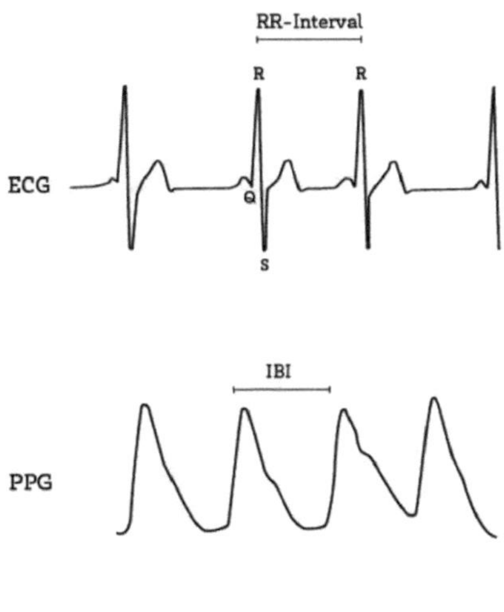

Figure 1.
ECG and PPG signals.

As previously discussed, both ECG and PPG system are able to provide information on cardiovascular activities. While ECG system allow better depiction of real cardiac movement through the measurement of the electrical signals produced by the action potential of the tissue, PPG allow adequate cardiovascular measurements such as heart rate and cardiac output only through pulsatile flow of blood in the arteries. Several studies have shown that the cardiovascular parameters collected through PPG systems are highly correlative and comparable to the measurements taken through standard ECG system [8, 10, 11]. This proves that despite not being able to illustrate exact cardiac waveforms or ectopic beats, PPG could serve as better alternative for portable heart monitoring device.

In terms of measurement accuracy, there are several factors to be considered to ensure the reliability of data collection. Topographical factor such as position of sensor placement on the body plays an important factor since different area of the body constitutes different accuracy of perfusion readings. The most accurate perfusion readings are recorded in earlobe; however, the wrist does allow perfusion readings with appropriate accuracy [9]. PPG watch is not subjected to electrical interference and drying or dropping-off of electrodes [8].

Therefore, this study proposes a PPG recording system for heart rate variability measurement that can be further used for mental stress assessment.

2. Method

ECG and PPG signal has been collected from 12 healthy subjects randomly selected with no prior symptoms of autonomic or cardiovascular disorder, ages between 20 and 30 years old. The data was collected with duration of 30 min including 10 min of adjustment, 10 min of rest (baseline) and 10 min of mental arithmetic testing. As a type of mental stress test, participants were needed to conduct an internet arithmetic test for 10 min in order to evaluate HRV under stress conditions such as time constraint. Lead II ECG setup with three electrodes were placed on the skin of the subject. For PPG signal, the wristband was placed on the left wrist. The subject was asked to sit down and make sure they are familiarized with the procedure. The ECG and PPG were recorded simultaneously after device was setup. The data was imported to the MATLAB software for the signal processing (**Figure 2**).

2.1 Signal processing

The recorded PPG and ECG signals were then pre-processed to extract the HRV using MATLAB software (**Figure 3**).

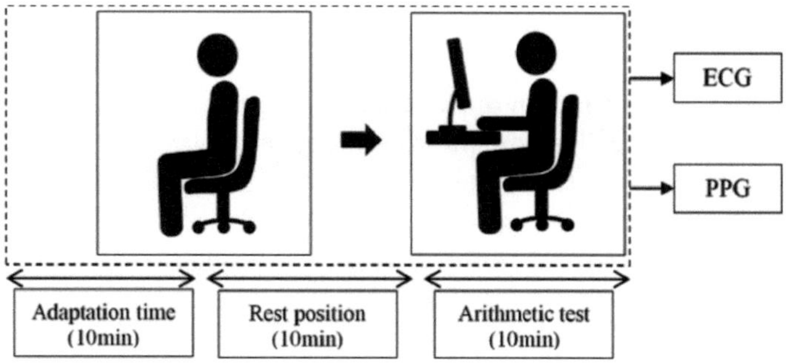

Figure 2.
Experiment setup.

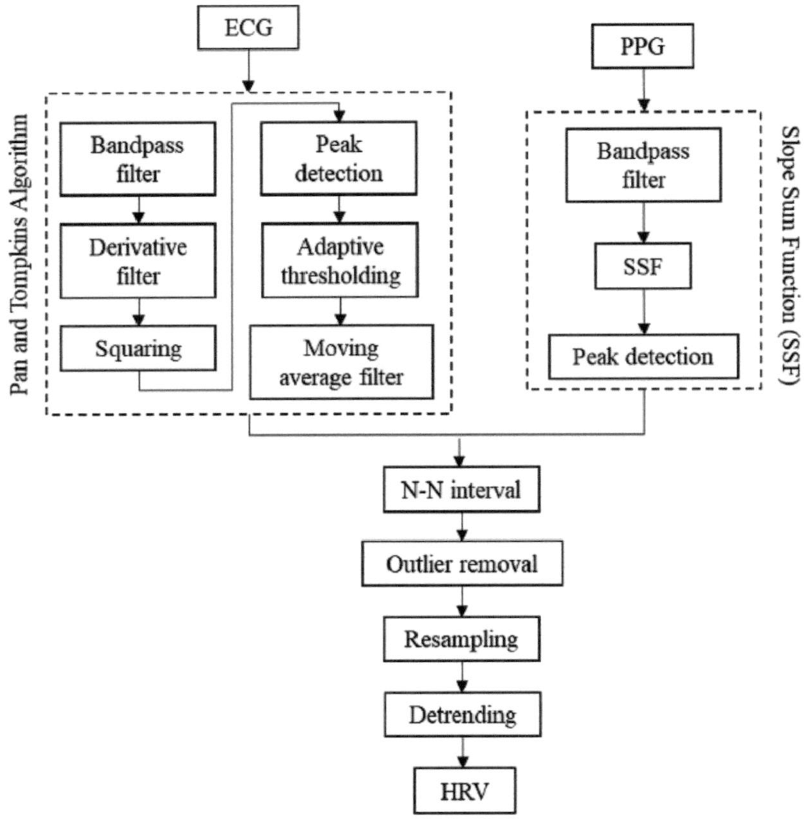

Figure 3.
Pan and Tompkins algorithm for ECG signal analysis and slope sum function (SSF) for PPG signal analysis.

2.1.1 HRV derived using PPG

The PPG signal began with the band pass filter to attenuate noises contained in the signals. The band-pass filter was made of cascaded lowpass and high-pass filters. The cut-off frequencies that have been used 5 and 11 Hz. The low pass filter (LPF) eliminates the noise from other part of body, such as the muscle noise and also 50 Hz power line noise. The high pass filter (HPF) which is used to remove the motion artifacts [12].

After that, the PPG signal undergo the slope sum function (SSF). This method is to enhance the systolic peak of the PPG pulse and to suppress the balance of the pressure waveform by using equation in Eq. (1) [13].

$$\mathrm{SSF} = \sum_{k=i-w}^{i} \Delta x_k, \quad \text{where } \Delta x_k = \begin{cases} \Delta S_k: & \Delta S_k > 0 \\ 0 \;\;: & \Delta S_k \leq 0 \end{cases} \tag{1}$$

where w and s_k are the length of the analyzing window and the filtered PPG signal, respectively. The SSF algorithm initialize the localization of the onset and offset of SSF then the pulse peak is identified as the local maxima within the range. The SSF signal produced coincides completely with the PPG pulse onset and offset and the pulse peaks appeared within the range of SSF pulse [14].

2.1.2 HRV derived using ECG

For ECG processing, Pan and Tompkins algorithm was implemented to get the HRV signal [15]. The Pan and Tompkins procedure are more complex as ECG signal contains superimposition of several waves (P, QRS, and T waves) as seen in **Figure 3** [16]. After initial denoising using BPF, the waveform undergoes differentiation process to obtain slope information overcome baseline drift. The next step is to perform signal squaring to emphasize higher frequency signal components (QRS waves) while attenuating components of low frequency. Resultant signal obtained through the squaring phase was then smoothed using moving average filter with a moving window integrator at 80 ms. A thresholding process is required to ensure that only the true QRS complex detected and the adaptive thresholds have been set for the classification of the locations of the detected R points.

The N-N interval was then computed and outliers presented in the signal was removed. Some of the data segment loss through the outlier extraction method was substituted by a new data segment using a linear interpolation method that resulted in NN intervals with nonequivalent moment sampling. However, the use of irregularly sampled NN intervals during HRV analysis characteristics such as frequency and TF analysis would cause generation of additional harmonic components and artifacts in (**Figure 4**) [16].

Therefore, the HRV signals were resampled at standard sampling frequency of 4 Hz [17]. Finally, the NN interval was passed through detrending process to overcome irregular trends.

2.2 HRV feature extraction

The following HRV features (**Table 1**) were computed based on the guidelines provided by Task Force of The European Society of Cardiology (ESC) [18].

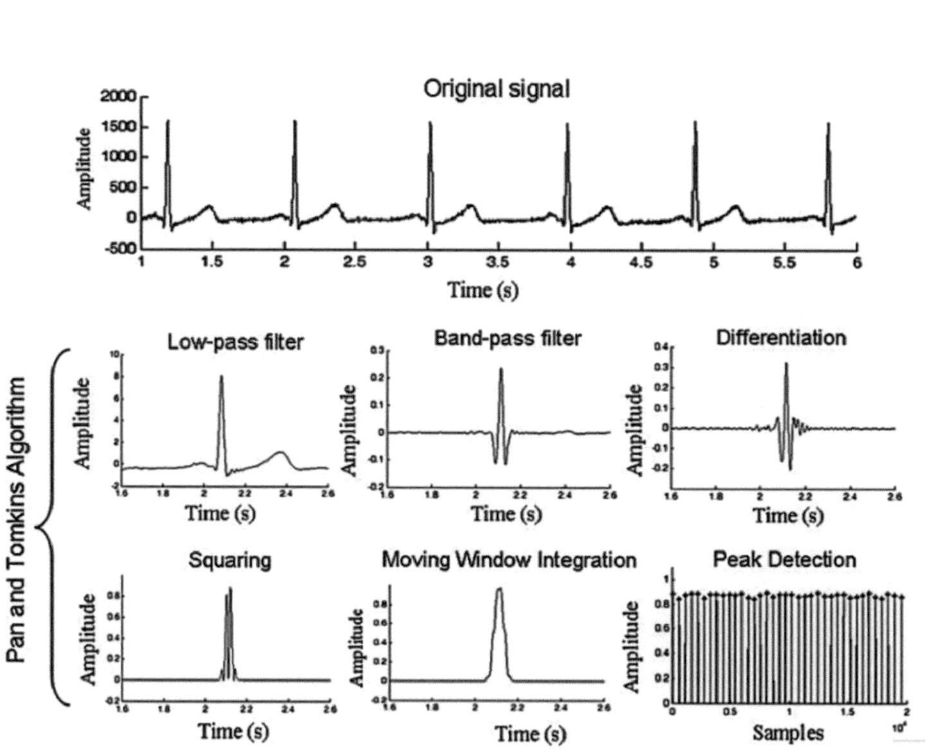

Figure 4.
Overview of HRV signal processing using Pan and Tompkins algorithm.

Processing method	HRV features	No. of features
Time analysis	HR, SDNN, SDANN, RMSSD, HTI, NN50, pNN50	7
Frequency analysis	VLF, LF, HF, LF/HF, LFnu, HFnu, TP	7
Nonlinear analysis	Shannon entropy: LF, HF, LF/HF, Total(O); Renyi entropy: LF, HF, LF/HF, Total(O)	8
	Total	22

Table 1.
Selected HRV features for extraction.

2.2.1 Time-domain features

In this study, the time domain has been analyzed from HRV signal. Besides that, HRV features were extracted which are standard deviation of the normal-to-normal intervals (SDNN), standard deviation of the average of normal-to-normal intervals (SDANN) and root mean square successive difference (RMSSD). SDNN, SDANN and RMSSD were calculated by using equations in Eq. (2), Eq. (3) and Eq. (4) respectively.

$$SDNN = \sqrt{\frac{1}{N-1}\sum_{n=1}^{N}[RR_n - mean(RR)]^2} \quad (2)$$

where N is total window length and NN is normal-to-normal time interval.

$$SDANN = \sqrt{\frac{1}{N_{s-1}}\sum_{n=1}^{N_s}[RR_n - mean(RR)]^2} \quad (3)$$

where N_5 is 5 min window length and NN is normal-to-normal time interval.

$$RMSSD = \sqrt{\frac{1}{N-2}\sum_{n=3}^{N}[I(n) - I(n-1)]^2} \quad (4)$$

where N is total window length.

2.2.2 Frequency-domain features

For this research, AR using the Burg estimation technique has been used to optimize forward and backward prediction errors. The power spectrum of the AR technique using the Burg estimation can be calculated as follows,

$$P_{Burg}(f) = \frac{\hat{e}_p}{\left|1 + \sum_{l=1}^{p}\hat{a}_p(l)e^{-2jfl}\right|^2} \quad (5)$$

where $\hat{e}_p$ represents the sum of both forward and backward prediction errors or the total least square error while p denotes the model order and $\hat{a}$ (l) indicate p_{th} order of the AR coefficient.

2.2.3 Nonlinear time-frequency features

Nonlinear analysis was performed using Modified B-distribution (MBD) as the technique is capable of providing high resolution TF distribution without cross-terms for HRV analysis. [16]. The kernel for the MBD as follows,

$$g(v,\tau) = \Gamma(\beta + j\pi\nu)^2/\Gamma^2(\beta) \quad (6)$$

where Γ defines as gamma function and β is a real positive number between 0 and 1 that regulates the trade-off between component resolution and cross-cutting elimination.

2.3 Multiscale HRV comparison and correlation analysis

In order to investigate the statistical significance ($p\text{-value} < 0.05$), Spearman's correlation is conducted between HRV features of multiple length under both resting and stress conditions. It is performed to determine the correlation between the HRV features produced through PPG signal in comparing with standardized ECG signal. A nonparametric Wilcoxon signed-rank test was performed to observe the difference between resting (baseline) and arithmetic stress test.

3. Result and discussion

This chapter presented the results obtained through pre-processing, feature extraction of HRV and multiscale comparison and correlation analysis along with relevant discussions of the findings.

3.1 Signal processing

The results of each pre-processing phase for HRV assessment and the resulting PPG signal HRV are shown in **Figures 5** and **6**, whereas the resulting ECG signal HRV is shown in **Figures 7** and **8**.

3.1.1 HRV derived using PPG

Figure 5 presented the attenuation of the PPG signal pulses after the application of the SSF conversion. The pulse peaks became more distinct throughout the entire signal duration using SSF conversion as lower ectopic beats were also amplified to

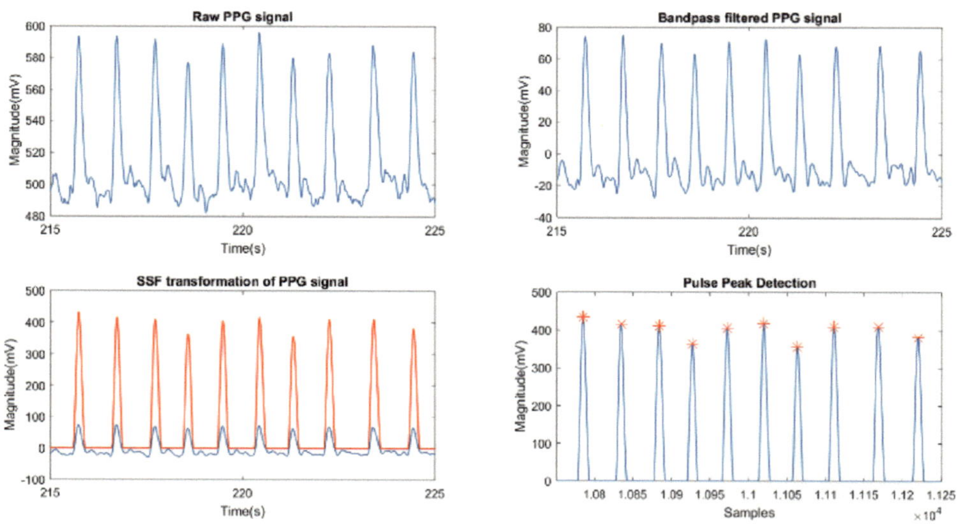

Figure 5.
The output of pulse peak detection from PPG signals using SSF.

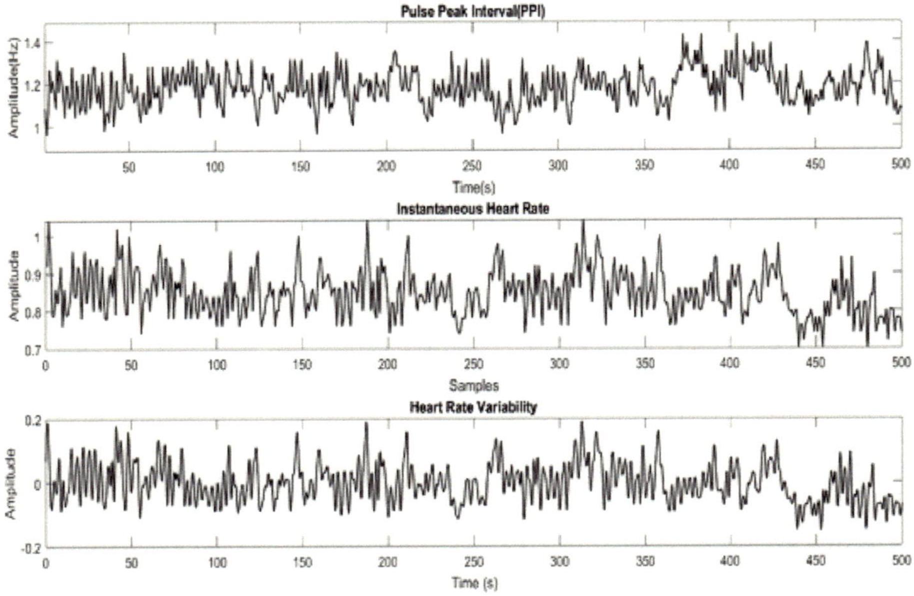

Figure 6.
The HRV signal obtained from pulse peak detected in SSF signal.

match ordinary pulse peaks that facilitate peak detection during thresholding method. **Figure 6** showed the resulting HRV signal that was obtained after removal, resampling and detrending of the outlier.

3.1.2 HRV derived using ECG

Figure 7 showed the changes in the ECG signal throughout the Pan and Tompkins algorithm processes. It can be seen that the algorithm was able to detect the R-R intervals throughout the signal excerpt. This method was chosen due to the simplicity and efficacy of this algorithm in QRS detection among adult subjects with 99.3% accuracy rate [15]. The subsequent HRV signal produced (illustrated in

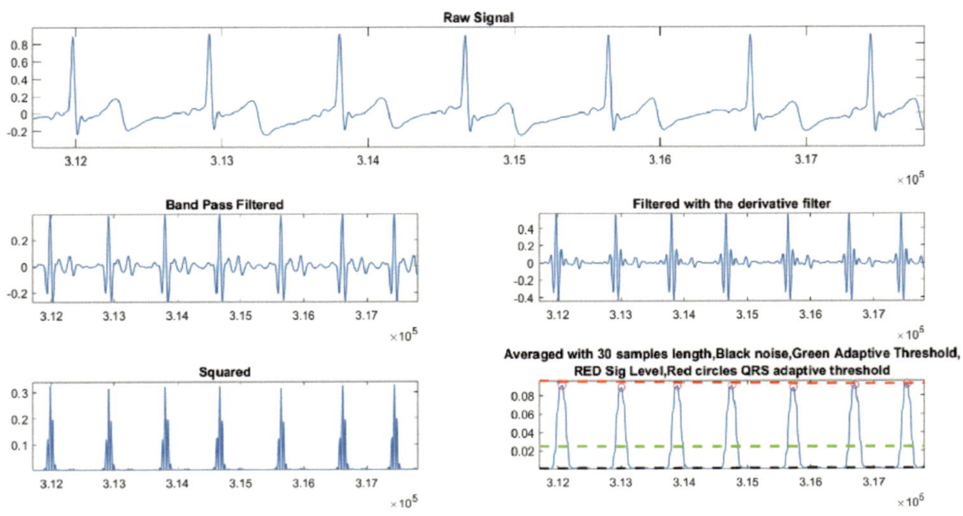

Figure 7.
The output of QRS peak detection from ECG signals using Pan and Tompkins algorithm.

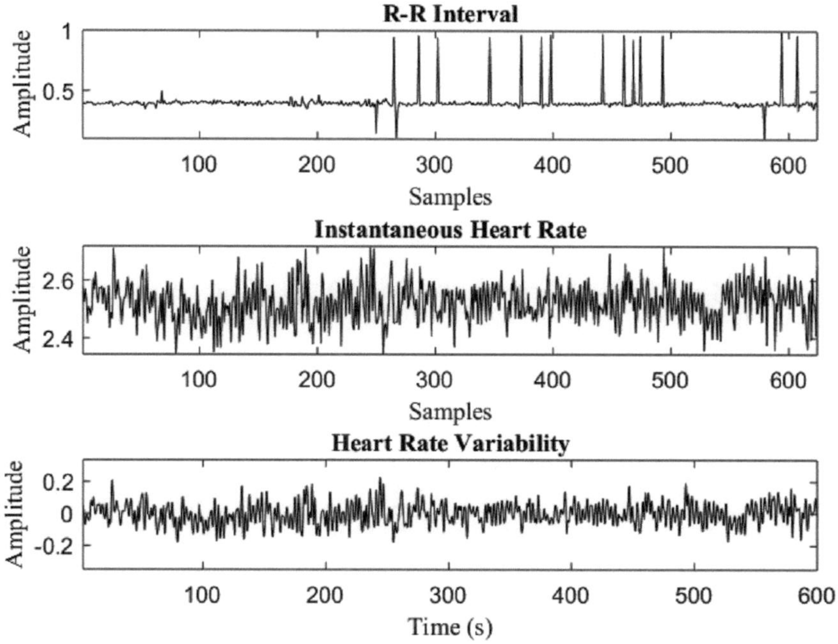

Figure 8.
The HRV signal obtained from RR peak detected.

Figure 8) after smoothing procedure was then used for feature analysis. Smoothing process which comprised of the outlier removal, resampling and detrending.

While evaluating the algorithm necessary for the HRV signal acquisition, it can be said that the pre-processing of the HRV signal recorded using the PPG system is simpler, as the signal contained only one type of wave (blood pulse) compared to the ECG signals usually containing a combination of three waves (P, QRS and T waves). Despite that, the HRV signal produced through both recordings do have relatively consistent magnitude.

3.2 HRV feature extraction

The analysis discussed in this section focuses mainly on the HRV features extracted using PPG method. Generally, the features selected have been associated with significant reactivity under stress conditions.

3.2.1 Time-domain features

The HRV signal obtained under resting and stress conditions were subsequently plotted in **Figure 9** which also showed the HRV obtained with time excerpts of 10 min duration. In addition, different lengths of HRV excerpts carry different

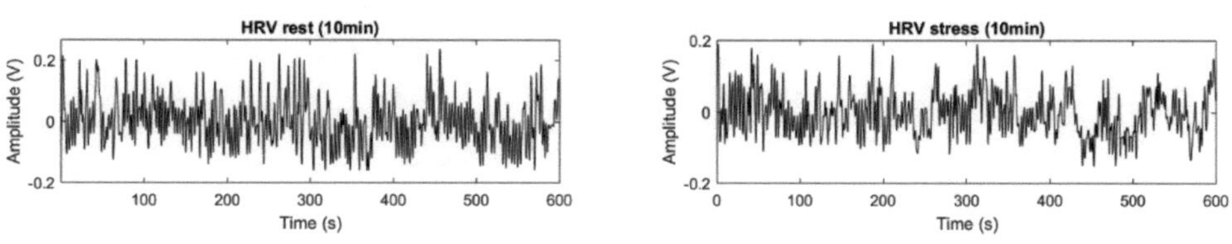

Figure 9.
Samples of PPG-derived HRV for 10 min from same sample between resting and stress condition.

weightage of information on the HRV of the sample. Longer HRV excerpts allow better visualization of fluctuations in the HRV measurements in both conditions. However, it is difficult to distinguish the difference of HRV changes between resting and stress testing through visual inspection only.

3.2.2 Frequency-domain features

The PSD can be classified into three components which are VLF band between 0.0033 and 0.04 Hz, LF band between 0.04 and 0.15 Hz and HF band between 0.15 and 0.4 Hz [18].

Based on the findings in **Figure 10**, the LF components increases during stress testing while HF components relatively decreases.

3.2.3 Nonlinear time-frequency features

For the plotted **Figure 11**, it was observed that more complex changes experienced during stress testing in 10 min. TFD plot was able to provide supplementary visualization of more complex changes within the HRV features during stress phase. Next, the changes within VLF and LF frequency bands were also more noticeable in TFD analysis.

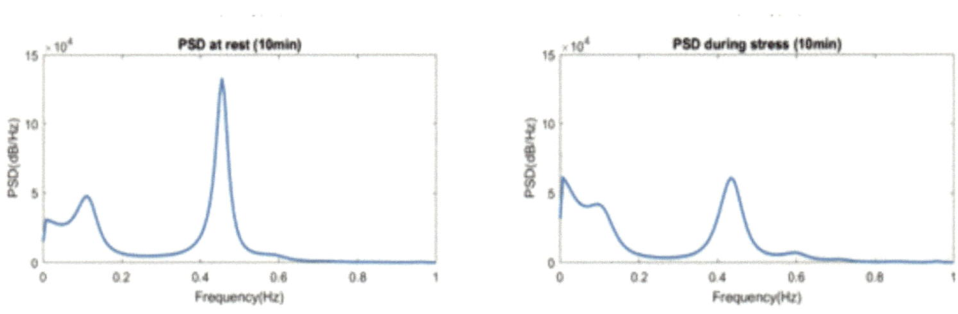

Figure 10.
Samples of PSD generated from PPG-derived HRV for 10 min from same sample between resting and stress condition.

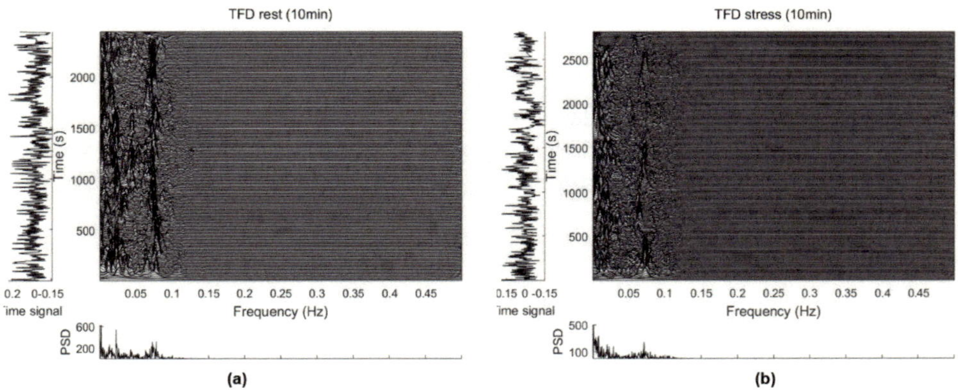

Figure 11.
Samples of TFD generated from PPG-derived HRV for 10 min from same sample between resting and stress condition.

3.3 Multiscale HRV comparison and correlation analysis

Based on this finding, it can be seen that most of the HRV features extracted using the PPG device produced similar measurements as the ECG, especially for the TA and FA features. However, for more sophisticated measurements, such as nonlinear TF characteristics, the correlation between the two techniques was less important, particularly for smaller HRV characteristics. This could be due to the fact that PPG waveform mainly reflects the central artery properties which means factors such as artery stiffness may attenuate the signal and resulted in differences of NN intervals obtained between different individuals [19]. The PPG signals are also influenced by other parasympathetic activity such as temperature variations

	Features	*r* Rest	*r* Stress
Time analysis	HR*	**0.964**	**0.970**
	SDNN*	**0.893**	**0.920**
	RMSSD*	**0.793**	**0.801**
	SDANN*	**0.909**	**0.964**
	NN50*	**0.659**	**0.907**
	pNN50*	**0.716**	**0.851**
	HTI*	**0.800**	**0.773**
Frequency analysis	VLF	**0.918**	0.491
	LF*	**0.773**	**0.764**
	HF*	**0.845**	**0.718**
	LF/HF*	**0.936**	**0.873**
	TP	**0.827**	0.364
	LFnu*	**0.936**	**0.873**
	Hfnu*	**0.936**	**0.864**
Nonlinear analysis	ShEn LF*	**0.800**	**0.818**
	ShEn HF*	**0.645**	**0.836**
	ShEn LFHF*	**0.727**	**0.818**
	ShEn O*	**0.709**	**0.773**
	ReEn LF	−0.064	0.255
	ReEn HF*	**0.873**	**0.909**
	ReEn LFHF*	**0.873**	**0.791**
	ReEn O*	**0.909**	**0.945**

In bold, Spearman's correlation coefficient (rho) greater than 0.6 and resulted correlation significant (prho < 0.05); and based on results, the time domain HRV features (except HTI) maintained a significantly high correlation coefficient. Frequency domain features at 10 minutes showed consistent significant correlation with the equivalent standard HRV features during both resting and stress phases. For non-linear analysis, Shanon Entropy measurements (ShEn LF, ShEn HF, ShEn LFHF and ShEn O) showed to be highly correlate with standard excerpt for HRV features at 10 minutes.
HR—mean of heart rate; SDNN—standard deviation of NN intervals; RMSSD—root mean square of the successive differences; SDANN—standard deviation of average NN intervals; NN50—NN intervals differing by more than 50 ms; pNN50—percentage of NN50 count; HTI—HRV triangular index; VLF—very low frequency; LF—low frequency; HF—high frequency; TP—total power; Lfnu—low frequency normalized unit; Hfnu—high frequency normalized unit; ShEn LF—Shannon entropy measurements; and ReEn—Renyi entropy measurements.
**Correlation is significant at the 0.01 level (2-tailed).*

Table 2.
Correlation between multi-length HRV features with standard of 10 min.

[20] and could significantly changes due to factors such as body age, vascular age, physical status, sleeping hours, physical activities [21].

Correlation analysis was performed to assess the interdependence between PPG-derived HRV and ECG-derived HRV as shown in **Table 2**.

In general, HRV features resulted less correlated in resting than during stress conditions. This is most likely due to the fact that HRV showed a more depressed dynamic during stress phase. Other than that, HRV features such as HR, NN50, TP, VLF, LF, HF, Lfnu, Hfnu, LF/HF, and Renyi entropy (LF, HF and Total(O)) has also showed significant correlation between the values measured for HRV excerpts collected using PPG and ECG. This prove that PPG is able to produce HRV signal with equivalent significant to HRV signal produced by ECG during stress testing [8, 22]. Besides, it can be deduced that HR, RMSSD, LF/HF, Lfnu and Hfnu features showed consistent characteristics as valid surrogate of the standard HRV which means regardless of length of HRV signal (between 1 and 10 min), these features would produce values that high correlate to value produced with standard HRV excerpt.

This study intends to investigate if there is different length of HRV excerpts provide valid measurement of HRV indices with comparison to standard 5-min excerpt for detection of mental stress. Although many studies have shown that HRV analyzes provide a reliable quantification technique for mental stress, it is hard to compare the precision of each method as their experimental design (i.e., duration of HRV characteristics) differs. Although it was claimed that the excerpt of 5-min HRV is the *gold standard* [18], the growing demand for wearable devices to instantly evaluate mental stress has increased interest in HRV computing characteristics shorter than the 5-min HRV standard [2]. In order to investigate the utility of various length of HRV excerpts in quantifying HRV features, 22 features were extracted at each time interval. The agreement between features at each time interval was compared with standard 5-min excerpt under both resting and stress phases. Overall, TA features (except HTI) conform significantly across all excerpts in correlation to standard excerpt while FA features (i.e., VLF, LF, HF, and TP) showed significant correlation across excerpts longer than 3 min while LFnu, HFnu and LF/HF showed consistent high correlation for all excerpts. As for time-frequency analysis, Shannon entropy measurements showed significant correlation for signal excerpts longer than 4 min while for Renyi entropy, only HF and Total(O) measurements showed significant correlation throughout all time excerpts.

Despite that, the limitation of these analyses is that correlation coefficient is blind to the possibility of bias caused by the difference in the mean or standard deviation between two measurements [23].

4. Conclusion

In comparison to conventional ECG, a correlation assessment between HRV characteristics obtained by PPG was also performed to observe any variation between the extracted measurements and analyze whether the PPG system is sufficiently robust to obtain HRV characteristics according to clinical standards. For this research, an ultra-short and short-term HRV feature was presumed to be a valid surrogate of the equivalent standard HRV if the feature sustained at a high correlation (i.e., rho > 0.6 and prho < 0.05) with the equivalent 5-min standard feature over all time scales and produced consistent trend and significant difference (p-value < 0.05) during the rest and stress phase. Therefore, it can be deduced that HR, RMSSD, LF/HF, LFnu and HFnu features showed consistent characteristics as valid surrogate of the standard HRV which means regardless of length of HRV signal (between 1 and 10 min), these features would produce values that high correlate to value produced with standard HRV excerpt. In the future, methods such as machine learning may be applied to test the accuracy between the use different PPG specifications such as measurement site, probe contact force and LED wavelengths which affect the reliability of its recordings or between different experimental protocol such as type of stressor and subject conditions.

Acknowledgements

This study was supported by the Research University Funding from Universiti Teknologi Malaysia, Malaysia (FRGS: R.J130000.7851.5F219).

Author details

Noor Aimie-Salleh*, Nurul Aliaa Abdul Ghani, Nurhafiezah Hasanudin and Siti Nur Shakiroh Shafie
School of Biomedical Engineering and Health Sciences, Faculty of Engineering, Universiti Teknologi Malaysia, Johor Bahru, Malaysia

*Address all correspondence to: aimie@biomedical.utm.my

IntechOpen

© 2019 The Author(s). Licensee IntechOpen. This chapter is distributed under the terms of the Creative Commons Attribution License (http://creativecommons.org/licenses/by/3.0), which permits unrestricted use, distribution, and reproduction in any medium, provided the original work is properly cited.

References

[1] Munla N, Khalil M, Shahin A, Mourad A. Driver stress level detection using HRV analysis. In: International Conference on Advances in Biomedical Engineering, ICABME. 2015. pp. 61-64

[2] Pecchia L, Castaldo R, Montesinos L, Melillo P. Are ultra-short heart rate variability features good surrogates of short-term ones? State-of-the-art review and recommendations. Healthcare Technology Letters. 2018;**5**(3):94-100. DOI: 10.1049/htl.2017.0090

[3] Thayer JF, Åhs F, Fredrikson M, Sollers JJ, Wager TD. A meta-analysis of heart rate variability and neuroimaging studies: Implications for heart rate variability as a marker of stress and health. Neuroscience & Biobehavioral Reviews. 2012;**36**(2):747-756. DOI: 10.1016/j.neubiorev.2011.11.009

[4] Mccraty R, Shaffer F. Heart rate variability: New perspectives on physiological mechanisms, assessment of self-regulatory capacity, and health risk. Global Advances in Health and Medicine. 2015;**4**(1):46-61. DOI: 10.7453/gahmj.2014.073

[5] Salahuddin L, Desok K. Detection of acute stress by heart rate variability using a prototype mobile ECG sensor. In: Proceedings - International Conference on Hybrid Information Technology, ICHIT, (November). 2006. pp. 1-25

[6] Bolanos M, Nazeran H, Haltiwanger E. Comparison of heart rate variability signal features derived from electrocardiography and photoplethysmography in healthy individuals. In: Annual International Conference of the IEEE Engineering in Medicine and Biology - Proceedings. 2006. pp. 4289-4294

[7] Peng RC, Zhou XL, Lin WH, Zhang YT. Extraction of heart rate variability from smartphone photoplethysmograms. Computational and Mathematical Methods in Medicine. 2015;**1**:1-11. DOI: 10.1155/2015/516826

[8] Selvaraj N, Jaryal A, Santhosh J, Deepak KK, Anand S. Assessment of heart rate variability derived from finger-tip photoplethysmography as compared to electrocardiography. Journal of Medical Engineering & Technology. 2008;**32**(6):479-484. DOI: 10.1080/03091900701781317

[9] Tamura T, Maeda Y, Sekine M, Yoshida M. Wearable photoplethysmographic sensors—Past and present. Electronics. 2014;**3**(2): 282-302. DOI: 10.3390/electronics3020282

[10] Lu G, Yang F, Taylor JA, Stein JF. A comparison of photoplethysmography and ECG recording to analyse heart rate variability in healthy subjects. Journal of Medical Engineering & Technology. 2009;**33**(8):634-641. DOI: 10.3109/03091900903150998

[11] Gil E, Orini M, Bailón R, Vergara JM, Mainardi L, Laguna P. Photoplethysmography pulse rate variability as a surrogate measurement of heart rate variability during non-stationary conditions. Physiological Measurement. 2010;**31**(9):1271-1290. DOI: 10.1088/0967-3334/31/9/015

[12] Awodeyi AE, Alty SR, Ghavami M. On the filtering of photoplethysmography signals. In: IEEE International Conference on Bioinformatics and Bioengineering. 2014. pp. 175-178

[13] Jang D, Park S, Hahn M, Park S. A real-time pulse peak detection algorithm for the photoplethysmogram. IEEE. 2014;**2**(1):45-49. DOI: 10.12720/ijeee.2.1.45-49

[14] Jang D-G, Farooq U, Park S-H, Hahn M. A robust method for pulse peak determination in a digital volume pulse waveform with a wandering baseline. IEEE Transactions on Biomedical Circuits and Systems. 2014; **8**(5):729-737. DOI: 10.1109/TBCAS.2013.2295102

[15] Pan J, Tompkins JW. A real-time QRS detection algorithm. IEEE Transactions on Biomedical Engineering. 1985;**32**:230-236. DOI: 10.1109/TBME.1985.325532

[16] Malarvili BK, Aimie S, Pranshul S. In: Ismail I, editor. Introduction to ECG Signal Processing using MATLAB. 1st ed. Johor Bahru, Malaysia: Penerbit UTM Press; 2018

[17] Chen H-K, Hu Y-F, Lin S-F. Methodological considerations in calculating heart rate variability based on wearable device heart rate samples. Computers in Biology and Medicine. 2018;**102**:396-401. DOI: 10.1016/j.compbiomed.2018.08.023

[18] Malik M, Thomas Bigger J, John Camm A, Kleiger RE, Malliani A, Moss AJ, et al. Heart rate variability: Standards of measurement, physiological interpretation, and clinical use. European Heart Journal. 1996;**17**: 354-381. DOI: 10.1093/oxfordjournals.eurheartj.a014868

[19] Elgendi M. On the analysis of fingertip photoplethysmogram signals. Current Cardiology Reviews. 2012;**8**(1): 14-25. DOI: 10.2174/157340312801215782

[20] Heyman F, Ahlberg N-E. Effect of rapid distension of large arteries and veins on the vascular tone of the fingers. Acta Medica Scandinavica. 2009;**183**(1–6):337-340. DOI: 10.1111/j.0954-6820.1968.tb10487.x

[21] Moraes J, Rocha M, Vasconcelos G, Vasconcelos Filho J, de Albuquerque V, Alexandria A. Advances in photopletysmography signal analysis for biomedical applications. Sensors. 2018; **18**(6):1894. DOI: 10.3390/s18061894

[22] Chang FC, Chang CK, Chiu C, Hsu SF, Lin YD. Variations of HRV analysis in different approaches. In: 2007 Computers in Cardiology; IEEE. 2007. pp. 17-20

[23] Castaldo R, Melillo P, Bracale U, Caserta M, Triassi M, Pecchia L. Acute mental stress assessment via short term HRV analysis in healthy adults: A systematic review with meta-analysis. Biomedical Signal Processing and Control. 2015, 2015;**18**:370-377. DOI: 10.1016/j.bspc.2015.02.012

Section 2

Heart Rate Variability and Hypothermia

Chapter 4

Modeling Thermoregulatory Responses to Cold Environments

Adam W. Potter, David P. Looney, Xiaojiang Xu, William R. Santee and Shankar Srinivasan

Abstract

The ability to model and simulate the rise and fall of core body temperature is of significant interest to a broad spectrum of organizations. These organizations include the military, as well as both public and private health and medical groups. To effectively use cold models, it is useful to understand the first principles of heat transfer within a given environment as well as have an understanding of the underlying physiology, including the thermoregulatory responses to various conditions and activities. The combination of both rational or first principles and empirical approaches to modeling allow for the development of practical models that can predict and simulate core body temperature changes for a given individual and ultimately provide protection from injury or death. The ability to predict these maximal potentials within complex and extreme environments is difficult. The present work outlines biomedical modeling techniques to simulate and predict cold-related injuries, and discusses current and legacy models and methods.

Keywords: hypothermia, cold injury, clothing, military, biophysics, survival

1. Introduction

Mitigating hot and cold injuries is a complex problem and has been shown to have significant links to a number of individualized factors, to include race, gender, job specialty, and geographic origin [1, 2]. There are many other individualized elements (e.g., fitness, body composition, and genetics) that are intuitively linked to these health outcomes; however, there is a lack of adequate data to scale that sufficiently addresses these issues.

The history of characterizing heat exchange and thermoregulatory functions in humans can be traced back to the late 1770s; where British military physiologist, Sir Charles Blagden conducted descriptive studies of man, dog, and beef steak responses in a hot room [3]. Mathematically describing heat exchange theory has roots in physics and with the development of the laws of thermodynamics and heat exchange, specifically as described in Fourier's law [4] a mathematical expression of the dynamics of heat balance in solids, simplified as:

$$\rho \cdot c \cdot \left(\frac{\partial T}{\partial t}\right) = \nabla k \ \ \nabla T + H \tag{1}$$

where ρ is density (g/m^3), c is specific heat [(kcal/°K. kg)], k is heat conductance [kcal/(hr cm °K)], T is temperature (°K), t is time (hours), and H is the net flow rate of heat other than by diffusion.

Key work by Pennes in 1948 [5], reported measured temperatures of tissue and blood at the forearm and enabled the creation of the bioheat transfer equation. This equation has proven to be a key underlying basis of future models, seen as:

$$\nabla \cdot k \ \ \nabla T + q_p + q_m - W_{C_b}(T - T_a) \ = \ \rho c_p\left(\frac{\partial T}{\partial t}\right) \tag{2}$$

where k (w/m °C) is the tissue thermal conductivity, T is tissue temperature in °C, q_p (w/m^3) is energy deposition rate, q_m (w/m^3) is metabolism, W (kg/m^3/s) is local tissue blood perfusion rate, C_b (J/kg/°C) is specific blood heat, T_a (°C) is arterial temperature, ρ (kg/m^3) is the tissue density, and c_p (J/kg/°C) is the specific tissue heat.

Conceptually, heat exchange between the human and the environment was first described by Lefevre in 1911; where he characterized the human as a sphere with an internal core that exchanged heat through the shell into the environment [6]. In 1934, Burton applied Fourier's law, presenting this exchange mathematically and describing the human as one uniform cylinder in what is considered by many as the first visual conceptualization of human thermoregulatory modeling [7].

Representation of the human in a thermoregulatory model is most often done by sectioning the human into nodes, segments, and elements; typically using one of four different designs, (1) one-node, (2) two-node, (3) multi-node, or (4) multi-element [8]. An example of the difference between these designs is shown in **Figure 1**; while the multi-element approach is more realistic human shape (e.g., finite analysis distribution). Typically each node represents an independent layer with unique thermal properties, each segment represents a section or grouped section of an area of the body, and each element represents multiple thermal components that make up the whole body (often more geometrically accurate to the shape of the human).

One node models are essentially empirically derived and do not include elements within the thermoregulatory response system. There are several one node thermoregulatory models that have been used extensively over time to predict core body temperature and thermal discomfort within a given environment [9–12].

Simple two-node models describe specific thermodynamic responses of a single segment, typically separated into concentric core and shell nodes. They have often been used examine thermal discomfort and physiological responses, to include the work by Gagge and Nishi [13–15], and several others [16–19]. Two node model approaches have been used where the two node design was applied to multi-segments [20–23]. Multi-node models are essentially expanded versions of the two-node methods with additional shells or layers within them where the heat balance is

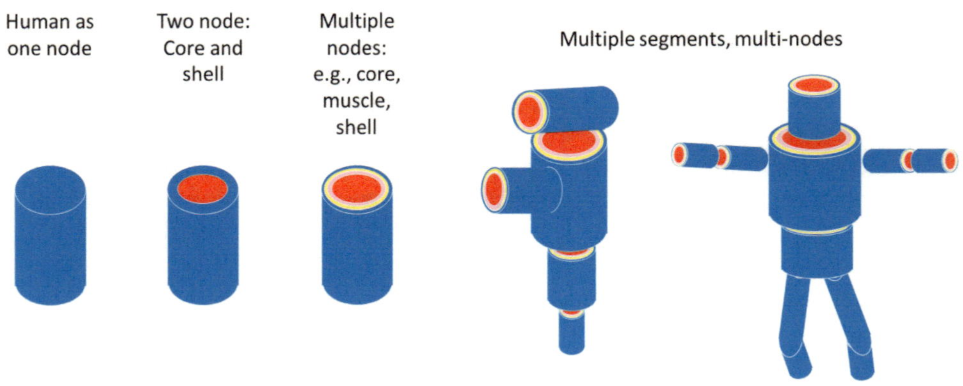

Figure 1.
Example of model designs.

calculated for each layer. Multi-node models, with both single- and multi-segment designs have become the more prevalent approach. The first multi-node model was developed by Crosbie et al. [24] and has been followed by many since [25–29]. Notable is the work of Solwijk and Hardy [30–33], where they first introduced the concepts of temperature set points and negative feedback in a controlled theory design. Their work has been built upon by many researchers over time [34–42]. The first multi-element model was originally published in 1961 by Wissler, and later improved upon [43–45]. Additional multi-element models include work by Smith [46], with the first three-dimensional (3D) transient multi-element model. As computation methods improved, a series of improvements has led to more realistic and complex models [8, 47–50].

While the majority of these models were developed with the intent of characterizing thermoregulation in various environments; several have been designed specifically to address cold environments or thermoregulatory events that specifically address cold issues (e.g., finger, hand, foot temperatures). With the intricacies of human response to cold, studies have focused on extremities, the specific areas most subject to cold injuries. One of the first attempts was by Molnar in 1957, used a heat balance approach to study hand temperature responses to cold [51]. This work was followed by work focused on finger freezing points [52–57] and whole hand modeling [58, 59]. Specific models have also been developed of the foot [60], toes [61], and facial tissues [62, 63]. Cold survival models have been developed over time to make predictions in both open air and submerged environments [64–68].

2. Clinical definitions of cold injuries

Characterizing cold related injuries is fairly complex, as the responses to cold have higher individual variability when compared to heat related injuries. From a clinical perspective, cold related injuries can be broadly divided into three categories: frostbite, nonfreezing cold injuries, and hypothermia. In addition, each of these has varying levels of severity and subcategories associated to them.

Frostbite is below the point at which skin tissue begins to freeze. While 0°C (32°F) is traditionally considered the freezing point of water, the freezing point of skin is understood to be marginally lower due to electrolytes [69]. Observed freezing points range from as low as −4.8°C to as high as −0.6°C [69, 70].

Nonfreezing cold injuries include an array of injury events where tissue freezing has not occurred but damage occurs. The level of severity of nonfreezing injuries is determined by the temperature, duration, and wetness of the exposure to the tissue. Four of the more common specific types of nonfreezing injuries include immersion (trench) foot, chilblain, cold urticaria, and cold-induced bronchoconstriction [71].

Immersion foot is a nonfreezing injury. The foot presents swollen, the skin is red initially but as severity increases the skin becomes lower in oxygen saturation and becomes cyanotic (purple, bluish discoloration) [69, 71]. Immersion foot is most often reported after tissue have been exposed for extended periods of time to non-freezing temperatures, between 0 and 15°C (32–60°F) [71]. The term ‘immersion’ itself refers to when the foot is actually immersed in water when the foot is wet within boots for sustained periods of time [69, 71].

Chilblain is a fairly common nonfreezing injury to the skin. It can occur during 1–5 hours of temperatures below 16°C (60°F) [69]. Cold urticaria is expressed as a quick onset of redness, swelling and itchiness of the skin in response to short-term exposure (i.e., minutes) to cold environments [71]. Cold-induced bronchoconstriction is a physiological response where an individual’s airways are narrowed during exercise in cold environments [69, 71–73].

44°C	Cell death (eggs cook)	Denaturing of protein
43°C		Heat stroke Hyperthermia >40°C
42°C		
41°C		
40°C	Heat stroke with neurological dysfunction	
39.5°C	High risk under compensable conditions [76]	Heat exhaustion
38.6°C	High risk under uncompensable conditions [76]	
37°C	Normal internal tempetature	Normal (Normothermia)
36°C		
35°C	Maximal shivering; increased blood pressure	Mild Hypothermia
34°C	Amnesia; dysarthria; cognitive impairment	
33°C	Ataxia; apathy	
32°C	Stupor	Moderate hypothermia
31°C	Shivering ceases; pupils dilate	
30°C	Cardiac arrhythmias; decreased cardiac output	
29°C	Unconsciousness	
28°C	Ventricular fibrillation likely; hypoventilation	Severe Hypothermia
27°C	Loss of reflexes and voluntary motion	
26°C	Acid-base disturbances; no response to pain	
25°C	Reduced cerebral blood flow	
24°C	Hypotension; bradycardia; pulmonary edema	
23°C	No corneal reflexes; areflexia	
19°C	Electroencephalographic silence	
18°C	Asytole	
15.2°C	Lowest recorded infant survival	
13.7°C	Lowest recorded adult survival	

Figure 2.
The range of human core temperatures and associated physiological responses [76].

Hypothermia is a broad category of cold injury and is clinically described to be the point at which core body temperature has dropped below 35°C (95°F) [74]. However, hypothermia is more specifically defined with four levels of severity; where normothermia (normal temperature level) is approximately 37°C (98.6°F), mild hypothermia is between 91.4–95°C (33–35°F), moderate hypothermia being 85.2–89.6°C (29–32°F), and severe hypothermia being 56.7–82.4°C (13.7–28°F) [69, 71]. **Figure 2** outlines specific core temperature reference points associated with physiological responses using work by Castellani et al. [69] and Pozos and Danzl [74] and described in Army Guidance [75].

3. Basics of thermophysiology

The human body is capable of maintaining thermal balance while operating within a wide range of temperatures. The human system generally maintains an internal core temperature (T_c) of approximately 37°C. Due to natural circadian rhythm, T_c fluctuates ~0.5°C daily. However, T_c can fluctuate based on physical activity or environmental conditions, and may range from 36.0–40.0°C. The microenvironment created between human skin and clothing typically must remain within 28–30°C to maintain thermal homeostasis at rest [45]. This microenvironment changes significantly with physical activity due to metabolic heat production and air movement.

Humans have an internal control system, primarily the preoptic area of the anterior hypothalamus, responsible for maintaining healthy body temperature. The hypothalamus uses feedback from two main sources, the skin and the blood. When temperature changes (hot or cold) are identified by either of these two sources, impulses are sent to the hypothalamus which in turn directs physiological changes to compensate for these temperatures. To protect from cold or heat injury, the human body attempts to either generate or dissipate heat to stay warm or cool off. Heat production is a natural process for humans and is a function of metabolism, oxidation of foods, and muscular activity. Heat transfer between the human and environment occurs via four pathways: conduction, convection, radiation, and evaporation. This heat exchange process is typically referred to as heat or thermal energy balance, and can be described in the heat balance equation:

$$S = M \pm W \pm R \pm C \pm K - E \,[\mathrm{W/m^2}] \tag{3}$$

where S is heat storage; M is metabolic rate; W is work rate; R is radiation; C is convection; K is conduction; and E is evaporation. Radiation is heat that is transferred via electromagnetic waves (e.g., solar radiation). Conduction is heat transfer due to the body's direct contact with a solid object (e.g., touching a cold surface). Convection is heat transfer between the body and a fluid such as air or water. Evaporation is heat loss to the environment due to the phase change from liquid to vapor, typically associated with evaporation of sweat and respiratory water.

Hyperthermia is when heat gain exceeds heat loss; while hypothermia occurs when body temperature drops below normal levels as heat production is inadequate to compensate for the rate of heat loss to the environment [77].

Vasoconstriction and vasodilation are the two key physiological responses of how heat transfer is regulated from the body to the periphery [78, 79]. Vasoconstriction is the constriction of blood vessels and occurs in response to cold environments to reduce the amount of blood flow to the skin. Vasoconstriction protects the internal organs from cold exposure but increases cold injury risk in the extremities due to lower blood flow and lower skin temperatures. Vasoconstriction in effect creates a two-layer distribution of body temperature; a cold outer shell surrounding a warmer core. The colder outer shell reduces heat loss to the environment by reducing the temperature gradient between the skin surface and the environment, and a colder surface radiates less heat.

Vasodilation is essentially the opposite of vasoconstriction; where blood vessels open to allow increased blood flow across the body and out to the extremities to enable increased heat dissipation [78, 79]. During these responses, there are other associated physiological responses that help compensate for the increased skin blood flow (e.g., increased heart rate and cardiac output).

The extremities are more affected by cold exposure than other parts of the body. When the human body cools, blood flow is reduced to the extremities (i.e., the hands and feet) decreasing the amount of warm blood flowing to these areas. It is a challenge to protect the hands and feet as they have lower metabolic heat production of the hands and feet due to their inherently small muscle mass and large surface area to mass ratio.

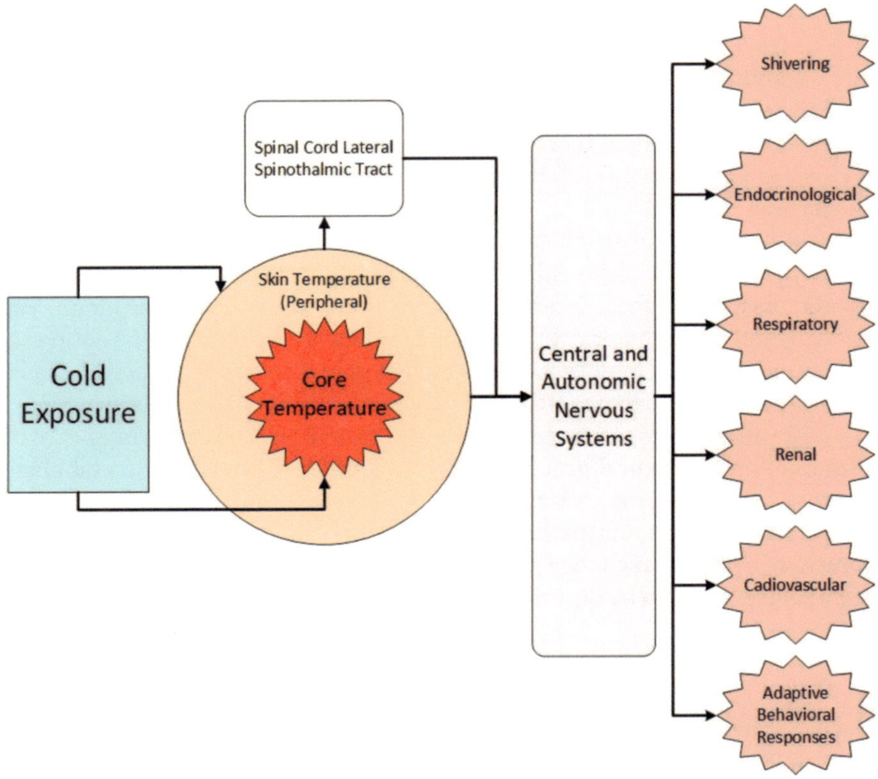

Figure 3.
Peripheral (skin) and core temperature influence on central nervous system (CNS) and physiological outcomes.

From a functional perspective, the balance of control within the human system depends on the response to cold exposure and interaction between peripheral (skin) and core body temperatures with the central nervous system (CNS) and the various physiological responses (**Figure 3**); [74].

4. Importance of clothing

Clothing has long been used to provide protection from environmental elements (heat, cold, etc.) or physical or biological hazards (e.g., rocks, thorns). Clothing properties and requirements vary widely among users and use cases. A single clothing ensemble cannot protect an individual from the extremes of the temperature spectrum of earth, being approximately −89°C at its coldest and 58°C at its warmest. However, clothing is a toll to protect each end of this spectrum of environmental extremes [80]. However, protections must be based on use cases to achieve the desired thermal comfort. For example, protective equipment for American football players (i.e., pads and helmet) is vastly different than protective equipment worn by soldiers (i.e., body armor, ballistic helmet). It should be noted that added protection may increase the thermal burden to wearers, and thus increases risk of heat injuries [81–83].

It is critical to understand the clothing option tradespace in order to predict and prepare for the impact clothing has on protecting or impairing human health. That is to say, the selection of the proper clothing, requires an understanding of how the human (physiology, anthropology, etc.), the anticipated activities (i.e., work rate, length of exposure and metabolic heat production), the work environments (temperature, humidity, etc.), and the biophysical properties of clothing worn (heat transfer performance) will interact in each workplace scenario.

4.1 Clothing biophysics

Clothing protects the wearer from environmental threats, but may impose a level of thermal burden. Both the biophysical resistances (thermal and evaporative) and spectrophotometric (reflectance, absorptivity, and transmittance) properties of clothing can have a significant influence on the impact of the environment on the wearer. Measurements of the biophysical properties of clothing can be used to model the impacts on thermal sensation (e.g., thermal comfort) and thermoregulatory responses (e.g., heat strain, cold protection). The thermal and evaporative resistances, wind effects, and spectrophotometric properties of the clothing are critical measurements for this purpose.

4.1.1 Thermal and evaporative resistance

Sweating thermal manikins have long been used to provide biophysical measures of clothing and equipment worn by the human [84]. While direct biophysical comparisons can be helpful, i.e., comparing one ensemble's value to another [85], a more informative approach is to combine these measured values with thermoregulatory modeling. Models enable the prediction of thermoregulatory responses based on different individuals, as well as varied environments, clothing, or activity levels.

The current standard for thermal manikin testing calls for two fundamental measures: thermal resistance (R_t) [86] and evaporative resistance (R_{et}) [87]. These two measures represent the dry heat exchange (R_t: convection, conduction, and radiation) and wet heat exchange (R_{et}: evaporation). After converting both R_t and R_{et} into units of clo and i_m [88, 89], a ratio can be used to describe an ensemble's evaporative potential (i_m/clo) [90].

Each ensemble should be tested using chamber conditions from the American Society for Testing and Materials (ASTM) standards for assessing R_t (ASTM F1291-16) and R_{et} (ASTM F2370-16) [86, 87] (**Table 1**).

Thermal resistance (R_t) is the dry heat transfer from the surface of the manikin through the clothing and into the environment, mainly from convection, and described as:

$$R_t = \frac{(T_s - T_a)}{Q/A}[\mathrm{m^2 K/W}] \tag{4}$$

where T_s is surface temperature and T_a is the air temperature, both in °C or °K. Q is power input (W) to maintain the surface (skin) temperature (T_s) of the manikin at a given set point; A is the surface area of the measurement in m^2. These measures of R_t can then be converted to units of clo:

$$1\,clo = 6.45(I_T) \tag{5}$$

Variable (unit)	Skin/surface temperature (T_s, °C)	Ambient temperature (T_a, °C)	Relative humidity (RH, %)	Wind velocity (V, ms^{-1})	Saturation (%)
R_t (m^2 K/W)	35	20	50	0.4	0
R_{et} (m^2 Pa/W)	35	35	40	0.4	100

Table 1.
*American Society for Testing and Materials standard chamber and manikin conditions for testing thermal (**R_t**) and evaporative (**R_{et}**) resistance.*

where I_T is the total insulation including boundary air layers. Evaporative resistance (R_{et}) is heat loss from the body in isothermal conditions ($T_s \approx T_a$), described as:

$$R_{et} = \frac{(P_{sat} - P_a)}{Q/A}[\mathrm{m^2\,Pa/W}] \quad (6)$$

where P_{sat} is vapor pressure in Pascal at the surface of the manikin (assumed to be fully saturated), and P_a is ambient vapor pressure, in Pascal, of the chamber environment. Measures of R_{et} can then be converted to a vapor permeability index (i_m), a non-dimensional measure of water vapor resistance of materials defined as:

$$i_m = \frac{60.6515 \frac{Pa}{{}^\circ C} R_t}{R_{et}} \quad (7)$$

4.1.2 Wind effects on thermal and evaporative resistance

In order to use the biophysical measures, i.e., measures of R_t (clo) and R_{et} (i_m) for thermoregulatory modeling there is a need to first estimate the effects of wind velocity on the biophysical characteristics of the ensemble (i.e., to determine how wind affects clo and i_m values). These effects are typically referred to as wind velocity coefficients or gamma values (^{g}) [91]. Historically, these coefficients were determined by collecting measurements of both R_t and R_{et} at multiple wind velocities above the ASTM standard of 0.4 m/s. However, recent work suggests these coefficient values can be accurately estimated from single wind velocity tests [91, 92].

Clothing properties and wind coefficients are critical inputs to a number of predictive mathematical models [10, 11, 93, 94], as they use these values to describe wind-related effects, such as intrinsic insulation (I_{cl}) and intrinsic permeability index (i_{cl}) for either the whole body or segments of the body, as seen with:

$$I_{cl} = I_t - \left(\frac{I_a}{f_{cl}}\right) \quad (8)$$

where I_a is insulation measured on a nude thermal manikin, I_t is total insulation, and (f_{cl}) is clothing area factor, calculated by:

$$f_{cl} = \frac{A}{A_{cl}} \quad (9)$$

where A (m^2) is surface area of the nude manikin, and A_{cl} (m^2) is surface area the clothed manikin.

True measures of A_{cl} require a three-dimensional scan. However, methods for estimating A_{cl} have been derived by McCullough et al. [95]. Simplified or estimated A_{cl} and f_{cl} is often used where a value of 1 is assumed for warm-weather or indoor clothing. For cold-weather clothing a value would be calculated from:

$$f_{cl} = 1.0 + 0.3 \cdot I_{cl} \quad (10)$$

While these estimation methods have been studied and produce acceptable variance between estimated and direct measured results [96], there are questions whether estimates remain acceptable for clothing insulation outside typical cold weather clothing insulation ranges, e.g., 0.2–1.7 clo [97].

Most clothing-based thermal models, by design, predict human thermoregulatory responses to various environmental conditions and therefore require quantitative insights into the change in clothing properties with changes in wind velocity. Furthermore, elements of wind can significantly influence physiological responses and injury outcomes in cold environments due to wind chill effects [69, 98, 99]. There has been work to develop that relates exposure time to predicted injury (e.g., frostbite) likely to occur due to temperature and levels of wind speed exposure [98].

5. Modeling risk and predicting heat and cold related injuries

Mathematical models can predict the human thermal response (e.g., metabolic heat production, core body temperature (T_c), endurance time) resulting from activity, environment, and clothing. These mathematical models are typically binned into one of three categories, either as rational, empirical, or hybrid. Rational (mechanistic) models mathematically represent phenomena based on an understanding of physics and physiology (biology, chemistry, physics). Empirical models mathematically reflect the observed relationship among experimental data. While both methods, rational and empirical, are scientifically valid approaches, perhaps the most effective approach is the hybrid or mixed model method that uses a combination of the two.

5.1 Rational models

Rational modeling incorporates equations that describe heat balance and thermoregulatory processes [100]. Two fundamental equations are used to describe internal heat balance and for heat exchange between skin and environment. One equation outlines the temperature gradient change from core to skin and can be seen as:

$$\rho c \cdot \frac{\partial T}{\partial t} = q_m + \lambda \cdot \nabla^2 T + \omega_{bl} \cdot \rho_{bl} c_{bl} \cdot (T_{bl} - T) \; [\mathrm{W\,m^{-3}}] \tag{11}$$

where ρ is tissue mass (kg m^{-3}), c is the specific heat of the tissue (kJ kg^{-1} °C^{-1}), T is the tissue temperature (°C), t is time (sec), q_m is metabolic heat production rate (W m^{-3}), λ is the tissue heat conductivity (W m^{-1} °C^{-1}), ∇^2 is a Laplace transform for heat conduction based on the tissue temperature gradient, ω_{bl} is blood flow rate (m^3 s^{-1} m^{-3} tissue), ρ_{bl} is blood flow mass (kg m^{-3}), c_{bl} is the blood specific heat (kJ kg^{-1} °C^{-1}), and T_{bl} is the blood temperature (°C).

The second equation describes heat exchange from the skin surface to the environment as:

$$-\lambda \cdot \frac{\partial T}{\partial n} = R + C + K + E \; [\mathrm{W\,m^{-2}}] \tag{12}$$

where λ is the tissue heat conductivity (W m^{-1} °C^{-1}), T is tissue temperature (°C), *n* is the tissue coordinate normal to the skin surface; while the balance is the array of avenues of heat exchange (W m^{-2}): R is radiative, C is convective, K is conductive, and E is evaporative.

Rational models of thermoregulatory processes usually include equations for the controlling signals of the thermoregulation system and equations for thermoregulatory actions such as sweating, vasodilation, vasoconstriction, and shivering.

Understanding the interplay between each of the different layers of the human (grossly consisting of core, muscle, fat, and skin) along with clothing and air layers within clothing is only the first step to modeling the human's response in a given environment. **Figure 4** shows the rational basis behind the SCENARIO model where the human is mathematically represented as one multi-layer cylinder, based on the relationship of the layers of the human, their respective physiological responses, and clothing [93, 94].

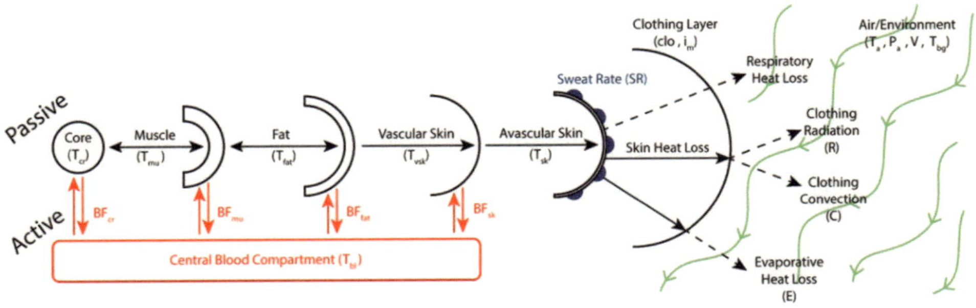

Figure 4.
Fundamental rational basis (SCENARIO model) [93], reused with permission. Note: BF_{cr} is core blood flow, BF_{mu} is muscle blood flow, BF_{fat} is muscle blood flow, BF_{sk} is skin blood flow.

5.2 Empirical models

Empirical models are mathematical representations of data, often using statistical methods such as regression or correlational analysis. An example model is the Heat Strain Decision Aid (HSDA), empirically derived by the U.S. Army from an extensive database of human studies that incorporates the biophysics of heat exchange [10, 11, 101] and predicts core temperature, maximum work times, sustainable work-rest cycles, water requirements, and the estimated likelihood of heat casualties. This model has been used to derive guidance and doctrine for military [102] and fluid intake guidance for the public [103]. The basis of HSDA includes both principles of heat exchange along with empirical predictions of physiological responses. Collectively 16 inputs from four elements (individual characteristics, physical activity, clothing biophysics, and environmental conditions) are used to mathematically predict the rise in core body temperature during physical activity [10].

5.3 Simple models

Originally developed by Holmér [104], a simple calculation was adopted by the International Organization Standardization (ISO) technical report (ISO 11079) [105], as an evaluation metric of the insulation required (IREQ) for given environments and activities to compare ensemble performance. The IREQ method functionally describes the concept for balancing the heat exchange between the human and the environment, and simplified as:

$$M - W = E_{res} + C_{res} + E + K + R + C + S \quad (13)$$

where M is metabolic heat produced, W is effective mechanical work and collectively M-W represents the heat produced within the human; while the opposite side of this balance, E_{res} and C_{res} represent the respiratory heat exchange (evaporative and convective), and E, K, R, and C represent the conventional heat exchange methods (evaporative, conductive, radiative, and convective) and S is heat storage.

The IREQ equation illustrates the rational balance between thermal insulation and heat transfer, seen as:

$$IREQ = \frac{\bar{t}_{sk} - t_{cl}}{R + C} \quad (14)$$

or more formally as:

$$IREQ = \frac{\bar{t}_{sk} - t_{cl}}{M - W - E_{res} - C_{res} - E} \tag{15}$$

where t_{sk} is mean skin temperature, t_{cl} clothing surface temperature, and $M - W - E_{res} - C_{res} - E = R + C$.

This method also determines the minimum and neutral IREQ (IREQmin and IREQneutral), and describes amounts of insulation needed to maintain thermal balance (minimum) and to maintain an equilibrium balance (neutral). The ISO 11079 also outlines general scenarios for the minimum required insulation ($IREQ_{min}$) for multiple work intensities and environments. Collectively this method provides a simple method for evaluating the effectiveness of specific cold weather clothing at protecting from cold injuries [106].

5.4 Key elements for model development

When developing a cold-based thermal model there are a number of physiological, environmental, and biophysical parameters that can and should be considered. Particular attention should be paid to the extremity temperatures blood flow and metabolic heat production.

5.4.1 Blood flow

As blood flow is a major component to the overall movement of heat, it is important to be able to predict blood flow to the muscle, skin, and distribution of blood flow to these regions within the body. **Table 2** outlines some historical methods used in models for predicting each of these elements.

5.4.2. Shivering

Shivering is where, in response to cold exposure, muscles involuntarily contract rhythmically off and on in an attempt to increase body temperature [74]. During

Prediction	Equation	Units	References
Cutaneous blood flow (q_s)	$q_s = q_{s,r} \cdot AVD \cdot CVCM \cdot CVCL \cdot CVCE$	mL 100 mL tissue^{-1} min^{-1}	[79, 107–115]
Skin vasodilation (dilat)	$dilat = \beta_{dil,1} \cdot error_1 + \beta_{dil,2} \cdot (warms - colds) + \beta_{dil,3} \cdot warm_1 \cdot warms$	L h^{-1}	[33]
Skin vasoconstriction (stric)	$stric = \beta_{str,1} \cdot error_1 + \beta_{str,2} \cdot (warms - colds) + \beta_{str,3} \cdot cold_1 \cdot colds$	L h^{-1}	[33]
Skin blood flow (bf_s)	$bf_s = 0.53 \cdot bf_{forearm} - 0.83$	mL min^{-1}	[116]
Local blood flow (lq_s)	$lq_s = \frac{q_{s,r} + \gamma_{dil} \cdot dilat}{1 + \gamma_{str} \cdot stric} \cdot Q_{10}^{\frac{T-T_0}{10}}$	L h^{-1}	[33]
Muscle blood flow (q_m)	$q_m = q_{m,r} + c_m \cdot \Delta M_w$	L h^{-1}	[33]
Muscle blood flow (bf_m)	$bf_m = 0.47 \cdot bf_{forearm} + 0.83$	mL min^{-1}	[116]

Note: q_s and $q_{s,r}$ are skin blood flow and rate; AVD is active vasodilation; CVC is cutaneous vascular conductance—addition of M (mediated), L (locally), and E (effect of exercise); β_{dil} and β_{str} are control coefficients for vasodilation and vasoconstriction; warms and colds refer to calculated net warm and cold receptors; bf forearm is blood flow at the forearm; γ dil and γ str are distribution coefficients for vasodilation and vasoconstriction; c_m is a proportionality coefficient; and M_W is metabolic heat produced from exercise.

Table 2.
Methods for predicting skin blood flow in thermoregulatory models.

Prediction	Equation	Units	References
Total shivering ($TOTM_{shiv}$)	$= 300 \cdot (T_h - T_{h,set}) + 1.35 \cdot \left(\sum_{m=1}^{14} W_{a,m} \cdot (\dot{q}_{s,m} - q_{s,set,m})\right) + 75 \cdot \left(\sum_{m=1}^{14} W_{a,m} \cdot (T_{s,m} - T_{s,set,m})\right)$	kcal h^{-1}	[37]
Maximal shivering ($Shiv_{max}$)	$= 30.5 + 0.348 \cdot VO_{2max} - 0.909 \cdot BMI - 0.233 \cdot age(yrs)$	$mLO_2\ kg^{-1}\ min^{-1}$	[37]
Metabolic rate of shivering (M_{shiv})	$= 60 \cdot (36.6 - T_{ty}) \cdot (34.1 - T_s)$	kcal h^{-1}	[30]
Metabolic rate of shivering (M_{shiv})	$= 36 \cdot (36.5 - T_{ty}) \cdot (32.2 - T_s) + 7 \cdot (32.2 - T_s)$	kcal h^{-1}	[117]
Metabolic rate to open air (M1)	$= 41.31 - 57.77 \cdot \frac{dT_s}{dt} - 5.01 \cdot (T_s - 34)$	W m^{-2}	[118]
Total metabolic rate (M2)	$= M1 + (894.15 - 23.79 \cdot T_{re})$	W m^{-2}	[118]
Total metabolic rate (M)	$= 0.0314 \cdot (T_s - 42.4) \cdot (T_{re} - 41.4)$	W kg^{-1}	[119]
Metabolic rate of shivering (M_{shiv})	$= \frac{155.5 \cdot (37 - T_{es}) + 47 \cdot (33 - T_s) - 1.57 \cdot (33 - T_s)^2}{\sqrt{BF\%}}$	W m^{-2}	[120]

Note: T is temperature; $_h$ is head; set is set point of temperatures; $W_{a,m}$ is a weighting coefficient; q_s is heat flux $_s$ is skin; BMI is body mass index; $_{ty}$ is Tympanic membrane; $_{re}$ is rectal; and $_{es}$ is esophageal; BF% is body fat percentage.

Table 3.
Methods for predicting shivering related model calculations.

Prediction	Equation	Units	References
Metabolic rate	$=1.44 + 1.94 \cdot S^{0.43} + 0.24 \cdot S^4$	W kg^{-1}	[124]
	$=3.5 + 6 \cdot S + 1.08 \cdot S \cdot G$	$mLO_2\ kg^{-1}\ min^{-1}$	[125]
	$=17.7 – 18.138 \cdot S + 9.72 \cdot S^2$	$mLO_2\ kg^{-1}\ min^{-1}$	[126]
	$=1.4 + 0.42 \cdot G + 3.68 \cdot S - 0.01 \cdot M - 0.03 \cdot Age$	W kg^{-1}	[127]
	$=1.5 \cdot M + 2 \cdot (M + L) \cdot (L \cdot M^{-1})^2 + \eta(M + L)(1.5 \cdot S^2 + 0.35 \cdot S \cdot G)$	W	[128]
	$=Ht \cdot (0.0136 \cdot Ht - 0.375)^{-1} \cdot (1.92 \cdot S^{0.176} - 1.445) \cdot Wt \cdot 10^5 \cdot (0.82 \cdot S^2 - 3.94 \cdot S + 9.66)$	$l\ O_2\ min^{-1}$	[129]

Note: G is grade (° for Ref. [125], % for others); Ht, height (inches for Ref. [129]); L, external load (kg); M, mass (kg); η, terrain factor; S, speed (mph for Ref. [129], m s^{-1} for others); VO_{2-rest}, resting oxygen consumption (ml kg^{-1} min^{-1}); Wt, weight (lbs).

Table 4.
Methods for predicting metabolic rates during walking or standing.

cold exposure the shivering response is a critical element to model, as the production of heat protects the body core temperature despite skin to the ambient heat loss. **Table 3** outlines some of the modeling approaches that have been used to predict the shivering response as they relate to the total metabolic rate (M) and the heat production from shivering (M_{shiv}).

5.4.3 Metabolic heat production

An individual's metabolic heat production can be estimated at rest and during activity using the assumed basal rate of 58.2 W/m^2 [121] and the estimated metabolic equivalents (METS) of activity; where 1 MET is resting. Ainsworth et al. [122] outlines a wide range of activities and their associated MET level for reference. However, there are metabolic rate estimation methods available based on energy costs of standing or walking (**Table 4**). Recently work has also been published that makes corrections to some of these prediction methods specific to traveling over snow terrain [123].

6. Summary and discussion

Mathematical models and decision aids are tools for inspiring advancements within the field of thermophysiology, and for providing solutions to help mitigate injury risk.

Scientifically based models have been used in the development of public [97, 98, 103, 104, 130–132] and military guidance [75, 131, 133], for forensic assessments [134–140], as well in the creation of operational tools for survival [141, 142]. Notably, the use of Xu and Werner's six cylinder model [41] was used to develop the Probability of Survival Decision Aid (PSDA), a computer model used to predict hypothermia and dehydration impact on functional time (i.e., duration of ability for useful work), and survival time while exposed to marine environments [67, 143, 144]. The PSDA model is underpinned by the rational principles described herein and the outputs are provided in a customized graphical user interface. This tool has been transitioned for use by Search and Rescue (SaR) personnel and continues to be refined and verified based on real-world feedback and data collected [144].

There is a need for continued advancement in the development of individualized modeling methods such as finite element models as well as providing models and decision aids that can be used in dynamic settings and for complex scenarios with prolonged durations. Additionally, inclusion of probabilistic and statistically based risk factors should be used as elements that help improve individualized predictions. The accessibility of the information from these tools continues to be a challenge for the scientific community. While providing usable information to the public, military, and other user communities should be the ultimate goal of these work efforts; feedback from these communities should be translated back to the scientists to ensure relevant improvements are made from real-world information.

Acknowledgements

This work is dedicated to the memory of Dr. Eugene H. Wissler (1927–2018). His pioneering efforts modeling heat transfer in the human body provided critical ground work that continues to be emulated by researchers.

The authors would also like to thank Dr. Scott Montain for oversight and review of this chapter.

Conflict of interest

The authors have no conflicts of interest to declare. Funding for this work has been provided by U.S. Army Medical Research and Materiel Command (USAMRMC), Military Operational Medicine Research Program (MOMRP).

Disclaimer

The views expressed in this paper are those of the authors and do not reflect the official policy of the Department of Army, Department of Defense, or the US Government.

Author details

Adam W. Potter[1,2*], David P. Looney[1], Xiaojiang Xu[1], William R. Santee[1,3] and Shankar Srinivasan[2]

1 Biophysics and Biomedical Modeling Division, United States Army Research Institute of Environmental Medicine, Natick, Massachusetts, United States of America

2 Rutgers University, School of Biomedical and Health Sciences, Newark, NJ, United States of America

3 Oak Ridge Institute for Science and Education (ORISE), Oak Ridge, TN, United States of America

*Address all correspondence to: adam.w.potter.civ@mail.mil

IntechOpen

© 2018 The Author(s). Licensee IntechOpen. This chapter is distributed under the terms of the Creative Commons Attribution License (http://creativecommons.org/licenses/by/3.0), which permits unrestricted use, distribution, and reproduction in any medium, provided the original work is properly cited.

References

[1] Carter IIIR, Cheuvront SN, Williams JO, Kolka MA, Stephenson LA, Sawka MN, et al. Epidemiology of hospitalizations and deaths from heat illness in soldiers. Medicine & Science in Sports & Exercise. 2005;**37**(8):1338-1344

[2] DeGroot DW, Castellani JW, Williams JO, Amoroso PJ. Epidemiology of US Army cold weather injuries, 1980-1999. Aviation, Space, and Environmental Medicine. 2003;**74**(5):564-570

[3] Blagden C. Experiments and observations in an heated room. Philosophical Transactions. 1775;**65**:111-123

[4] Fourier J. Theorie analytique de la chaleur, par M. Fourier. Paris, France: Chez Firmin Didot, père et fils; p. 1822

[5] Pennes HH. Analysis of tissue and arterial blood temperatures in the resting human forearm. Journal of Applied Physiology. 1948;**1**(2):93-122

[6] Lefevre J. Chaleur animale et bioénergétique. Paris, France: Masson et cie; 1911

[7] Burton AC. The application of the theory of heat flow to the study of energy metabolism: Five figures. The Journal of Nutrition. 1934;7(5):497-533

[8] Fu G. A transient, 3-D mathematical thermal model for the clothed human [Ph.D. dissertation]. Kansas State University, Manhattan, Kansas; 1995

[9] Givoni B, Goldman RF. Predicting metabolic energy cost. Journal of Applied Physiology. 1971;**30**(3):429-433

[10] Givoni B, Goldman RF. Predicting rectal temperature response to work, environment, and clothing. Journal of Applied Physiology. 1972;**32**(6):812-822

[11] Potter AW, Blanchard LA, Friedl KE, Cadarette BS, Hoyt RW. Mathematical prediction of core body temperature from environment, activity, and clothing: The heat strain decision aid (HSDA). Journal of Thermal Biology. 2017;**64**:78-85

[12] Fanger PO. Thermal comfort. Analysis and applications in environmental engineering. In: Thermal Comfort: Analysis and Applications in Environmental Engineering. Copenhaen, Denmark: Danish Technical Press; 1970

[13] Gagge AP, Stolwijk JA, Saltin B. Comfort and thermal sensations and associated physiological responses during exercise at various ambient temperatures. Environmental Research. 1969;**2**(3):209-229

[14] Gagge AP. An effective temperature scale based on a simple model of human physiological regulatory response. ASHRAE Transactions. 1971;**77**:247-262

[15] Nishi Y, Gagge AP. Humid operative temperature. A biophysical index of thermal sensation and discomfort. Journal de Physiologie. 1971;**63**(3):365

[16] Takada S, Kobayashi H, Matsushita T. Thermal model of human body fitted with individual characteristics of body temperature regulation. Building and Environment. 2009;**44**(3):463-470

[17] Azer NZ, Hsu S. The prediction of thermal sensation from a simple model of human physiological regulatory response. ASHRAE Transactions. 1977;**83**(Pt 1):88-102

[18] Jones BW, Ogawa Y. Transient response of the human-clothing system. Journal of Thermal Biology. 1993;**18**(5-6):413-416

[19] Kingma BR, Frijns AJ, Schellen L, van Marken Lichtenbelt WD. Beyond

the classic thermoneutral zone: Including thermal comfort. Temperature. 2014;**1**(2):142-149

[20] Kohri I, Mochida T. Evaluation method of thermal comfort in a vehicle with a dispersed two-node model part 1—Development of dispersed two-node model. Journal of the Human-Environment System. 2002;**6**(1):19-29

[21] Kaynakli O, Unver U, Kilic M. Evaluating thermal environments for sitting and standing posture. International Communications in Heat and Mass Transfer. 2003;**30**(8):1179-1188

[22] Kaynakli O, Kilic M. Investigation of indoor thermal comfort under transient conditions. Building and Environment. 2005;**40**(2):165-174

[23] Foda E, Sirén K. A new approach using the Pierce two-node model for different body parts. International Journal of Biometeorology. 2011;**55**(4):519-532

[24] Crosbie RJ, Hardy JD, Fessenden E. Electrical analog simulation of temperature regulation in man. IRE Transactions on Bio-Medical Electronics. 1961;**8**(4):245-252

[25] Fiala D, Lomas KJ, Stohrer M. A computer model of human thermoregulation for a wide range of environmental conditions: The passive system. Journal of Applied Physiology. 1999;**87**(5):1957-1972

[26] Fiala D, Lomas KJ, Stohrer M. Computer prediction of human thermoregulatory and temperature responses to a wide range of environmental conditions. International Journal of Biometeorology. 2001;**45**(3):143-159

[27] Kingma BR. Human thermoregulation; a synergy between physiology and mathematical modeling [doctoral dissertation]. Maastricht University

[28] Kingma BR, Frijns AJ, Saris WH, van Steenhoven AA, van Marken Lichtenbelt WD. Mathematical modeling of human thermoregulation: A neurophysiological approach to vasoconstriction. In: Computational Intelligence. Berlin, Heidelberg: Springer; 2012. pp. 307-316

[29] Davoodi F, Hasanzadeh H, Zolfaghari SA, Maerefat M. Developing a new individualized 3-node model for evaluating the effects of personal factors on thermal sensation. Journal of Thermal Biology. 2017;**69**:1-2

[30] Stolwijk JAJ, Hardy JD. Temperature regulation in man—A theoretical study. Pflügers Archiv für die Gesamte Physiologie des Menschen und der Tiere. 1966;**291**:129-162

[31] Stolwijk JAJ. A Mathematical Model of Physiological Temperature Regulation in Man. NASA-CR-1855. Washington, DC: National Aeronaucis and Space Administration; 1971

[32] Stolwijk JAJ. Mathematical models of thermal regulation. Annals of the New York Academy of Sciences. 1980;**335**(1):98-106

[33] Stolwijk JAJ, Hardy JD. "Control of Body Temperature". Comprehensive Physiology 2011, Supplement 26: Handbook of Physiology, Reactions to Environmental Agents. First Published in Print 1977. John Wiley & Sons, Inc.; 2010. pp. 45-68. DOI: 10.1002/cphy.cp090104

[34] Montgomery LD. A model of heat transfer in immersed man. Annals of Biomedical Engineering. 1974;**2**(1):19-46

[35] Montgomery LD. Biothermal simulation of scuba divers. Aviation, Space, and Environmental Medicine. 1975;**46**(6):814-818

[36] Kuznetz LH. Control of thermal balance by a liquid circulating garment based on a mathematical representation of the human thermoregulatory system [Ph.D. thesis]. Berkeley: California Univ.; 1976

[37] Gordon RG, Roemer RB, Horvath SM. A mathematical model of the human temperature regulatory system-transient cold exposure response. IEEE Transactions on Biomedical Engineering. 1976;**6**:434-444

[38] Tikuisis PE, Gonzalez RR, Pandolf KB. Thermoregulatory model for immersion of humans in cold water. Journal of Applied Physiology. 1988;**64**(2):719-727

[39] Tikuisis PE, Gonzalez RR, Pandolf KB. Prediction of human thermoregulatory responses and endurance time in water at 20 and 24 degrees C. Aviation, Space, and Environmental Medicine. 1988;**59**(8):742-748

[40] Werner J, Webb P. A six-cylinder model of human thermoregulation for general use on personal computers. The Annals of Physiological Anthropology. 1993;**12**:123-134

[41] Xu X, Werner J. A dynamic model of the human/clothing/environment-system. Applied Human Science: Journal of Physiological Anthropology. 1997;**16**(2):61-75

[42] Munir A, Takada S, Matsushita T. Re-evaluation of Stolwijk's 25-node human thermal model under thermal-transient conditions: Prediction of skin temperature in low-activity conditions. Building and Environment. 2009;**44**(9):1777-1787

[43] Wissler EH. Steady-state temperature distribution in man. Journal of Applied Physiology. 1961;**16**:734-740

[44] Wissler EH. A mathematical model of the human thermal system. The Bulletin of Mathematical Biophysics. 1964;**26**:147-166

[45] Wissler EH. In: Shitzer A, Eberhart R, editors. Mathematical Simulation of Human Thermal Behavior Using Whole Body Models. New York: Plenum Press, Heat Transfer in Medicine and Biology; 1985. pp. 325-373

[46] Smith CE. A transient 3-D model of the human thermal system [doctoral dissertation, PhD Thesis]. Kansas State University; 1991

[47] Ferreira MS, Yanagihara JI. A transient three-dimensional heat transfer model of the human body. International Communications in Heat and Mass Transfer. 2009;**36**(7):718-724

[48] Schwarz M, Krueger MW, Busch HJ, Benk C, Heilmann C. Model-based assessment of tissue perfusion and temperature in deep hypothermic patients. IEEE Transactions on Biomedical Engineering. 2010;**57**(7):1577-1586

[49] Sun X, Eckels S, Zheng ZC. An improved thermal model of the human body. HVAC&R Research. 2012;**18**(3):323-338

[50] Tang Y, He Y, Shao H, Ji C. Assessment of comfortable clothing thermal resistance using a multi-scale human thermoregulatory model. International Journal of Heat and Mass Transfer. 2016;**98**:568-583

[51] Molnar GW. Heat transfer through the hand. In: Fisher FR, editor. Protection and Functioning of the Hands in Cold Climates. Washington, DC: National Academy of Sciences; 1957:15-45

[52] Shitzer AV, Stroschein LA, Santee WR, Gonzalez RR, Pandolf KB. Quantification of conservative

endurance times in thermally insulated cold-stressed digits. Journal of Applied Physiology. 1991;**71**(6):2528-2535

[53] Shitzer AV, Stroschein LA, Gonzalez RR, Pandolf KB. Lumped-parameter tissue temperature-blood perfusion model of a cold-stressed fingertip. Journal of Applied Physiology. 1996;**80**(5):1829-1834

[54] Shitzer A, Stroschein LA, Vital P, Gonzalez RR, Pandolf KB. Numerical analysis of an extremity in a cold environment including countercurrent arterio-venous heat exchange. Journal of Biomechanical Engineering. 1997;**119**(2):179-186

[55] Tikuisis P, Ducharme MB. Finite-element solution of thermal conductivity of muscle during cold water immersion. Journal of Applied Physiology. 1991;**70**(6):2673-2681

[56] Ducharme MB, Tikuisis P. Forearm temperature profile during the transient phase of thermal stress. European Journal of Applied Physiology and Occupational Physiology. 1992;**64**(5):395-401

[57] Tikuisis P. Finger cooling during cold air exposure. Bulletin of the American Meteorological Society. 2004;**85**(5):717-724

[58] Montgomery LD, Williams BA. Effect of ambient temperature on the thermal profile of the human forearm, hand, and fingers. Annals of Biomedical Engineering. 1976;**4**(3):209-219

[59] Lotens WA. Simulation of hand cooling due to touching cold materials. European Journal of Applied Physiology and Occupational Physiology. 1992;**65**(1):59-65

[60] Lotens WA, Heus R, Van de Linde FJG. A 2-node thermoregulatory model for the foot. In: Mercer JB editor. Thermal physiology. Elsevier Science Publishers; 1989:769-775

[61] Xu X, Endrusick TL, Santee WB, Kolka MA. Simulation of toe thermal responses to cold exposure while wearing protective footwear. SAE Technical Paper. 2005

[62] Tikuisis P, Ducharme MB, Brajkovic D. Prediction of facial cooling while walking in cold wind. Computers in Biology and Medicine. 2007;**37**(9):1225-1231

[63] Tikuisis P, Osczevski RJ. Dynamic model of facial cooling. Journal of Applied Meteorology. 2002;**41**(12):1241-1246

[64] Tikuisis P. Predicting survival time for cold exposure. International Journal of Biometeorology. 1995;**39**(2):94-102

[65] Tikuisis P. Prediction of survival time at sea based on observed body cooling rates. Aviation, Space, and Environmental Medicine. 1997;**68**(5):441-448

[66] Xu X, Tikuisis P, Gonzalez R, Giesbrecht G. Thermoregulatory model for prediction of long-term cold exposure. Computers in Biology and Medicine. 2005;**35**(4):287-298

[67] Xu X, Turner CA, Santee WR. Survival time prediction in marine environments. Journal of Thermal Biology. 2011;**36**(6):340-345

[68] Xu X, Tikuisis P, Giesbrecht G. A mathematical model for human brain cooling during cold-water near-drowning. Journal of Applied Physiology. 1999;**86**(1):265-272

[69] Castellani JW, Young AJ, Ducharme MB, Giesbrecht GG, Glickman E, Sallis RE. Prevention of cold injuries during exercise. Medicine & Science in Sports & Exercise. 2006;**38**:2012-2029

[70] Keatinge WR, Cannon P. Freezing-point of human skin. Lancet. 1960;**I**:11-14

[71] Hamlet MP. Nonfreezing cold injuries. In: Auebach PS, editor. Textbook of Wilderness Medicine. St. Louis, MO: Mosby; 2001. pp. 129-134

[72] Evans TM, Rundell KW, Beck KC, Levine AM, Baumann JM. Cold air inhalation does not affect the severity of EIB after exercise or eucapnic voluntary hyperventilation. Medicine & Science in Sports & Exercise. 2005;**37**(4):544-549

[73] Wilber RL, Rundell KW, Szmedra L, Jenkinson DM, Im J, Drake SD. Incidence of exercise-induced bronchospasm in Olympic winter sport athletes. Medicine & Science in Sports & Exercise. 2000;**32**(4):732-737

[74] Pozos RS, Danzl DF. Human physiological responses to cold stress and hypothermia. In: Pandolf KB, Burr RE, editors. Textbooks of Military Medicine: Medical Aspects of Harsh Environments. Vol. 1. Falls Church, YA: Office of the Surgeon General, U. S. Army; 2002. pp. 351-382

[75] Department of the Army. Prevention and Management of Cold-Weather Injuries. Washington, DC: 2005. Report No.: TB MED 508

[76] Sawka MN, Latzka WA, Montain SJ, Cadarette BS, Kolka MA, Kraning KK, et al. Physiologic tolerance to uncompensable heat: Intermittent exercise, field vs laboratory. Medicine and Science in Sports and Exercise. 2001;**33**(3):422-430

[77] Sawka MN, Young AJ. Physiological systems and their responses to conditions of heat and cold. In: American College of Sports Medicine. ACSM's Advanced Exercise Physiology. Philadelphia, PA: Lippincott Williams & Wilkins; 2006. pp. 535-563

[78] Johnson JM, Niederberger MA, Rowell LB, Eisman MM, Brengelmann GL. Competition between cutaneous vasodilator and vasoconstrictor reflexes in man. Journal of Applied Physiology. 1973;**35**:798-803

[79] Charkoudian N. Skin blood flow in adult human thermoregulation: How it works, when it does not, and why. Mayo Clinic Proceedings. 2003;**78**(5):603-612

[80] Astrand A, Rodahl I. Textbook of Work Physiology. New York, NY: McGraw Hill; 1986. pp. 104-112

[81] Potter AW, Gonzalez JA, Karis AJ, Xu X. Biophysical assessment and predicted thermophysiologic effects of body armor. PLoS One. 2015;**10**(7):e0132698

[82] Potter AW, Karis AJ, Gonzalez JA. Biophysical characterization and predicted human thermal responses to US army body armor protection levels (BAPL). In: Technical Report, T13-5, ADA#585406. Natick, MA 01760 USA: U.S. Army Research Institute of Environmental Medicine ; 2013. Available from: www.dtic.mil/dtic/tr/fulltext/u2/a585406.pdf

[83] Larsen B, Netto K, Aisbett B. The effect of body armor on performance, thermal stress, and exertion: A critical review. Military Medicine. 2011;**176**(11):1265-1273

[84] Xu X, Gonzalez JA, Karis AJ, Rioux TP, Potter AW. Use of thermal mannequins for evaluation of heat stress imposed by personal protective equipment. In: Shiels B, Lehtonen K, editors. Performance of Protective Clothing and Equipment: 10th Volume, Risk Reduction through Research and Testing, ASTM STP1593. West Conshohocken, PA: ASTM International; 2016. pp. 286-296

[85] Potter AW, Gonzalez JA, Karis AJ, Santee WR, Rioux TP, Blanchard LA. Biophysical characteristics and measured wind effects of chemical protective ensembles with and without body armor. In: Technical Report,

T15-8; ADA#621169. Natick, MA, 01760, USA: US Army Research Institute of Environmental Medicine; 2015. Available from: www.dtic.mil/dtic/tr/fulltext/u2/a621169.pdf

[86] American Society of Testing and Materials International (ASTM). Standard Test Method for Measuring the Thermal Insulation of Clothing Using a Heated Manikin (ASTM F1291-16) [Standard]. Philadelphia, Pa.: ASTM; 2016

[87] American Society of Testing and Materials International (ASTM). Standard Test Method for Measuring the Evaporative Resistance of Clothing Using a Sweating Manikin (ASTM F2370-16) [Standard]. Philadelphia, Pa.: ASTM; 2016

[88] Gagge AP, Burton AC, Bazett HC. A practical system of units for the description of the heat exchange of man with his environment. Science. 1941;**94**:428-430

[89] Woodcock AH. Moisture transfer in textile systems, Part I. Textile Research Journal. 1962;**32**(8):628-633

[90] Woodcock AH. Moisture permeability index—A new index for describing evaporative heat transfer through fabric systems. In: Technical Report (TR-EP-149). Natick, MA 01702, USA: Quartermaster Research and Engineering Command; 1961

[91] Potter AW, Gonzalez JA, Karis AJ, Rioux TP, Blanchard LA, Xu X. Impact of estimating thermal manikin derived wind velocity coefficients on physiological modeling. In: Technical Report, ADA#607972. Natick, MA, 01760, USA: US Army Research Institute of Environmental Medicine; 2014. Available from: www.dtic.mil/dtic/tr/fulltext/u2/a607972.pdf

[92] Potter AW. Method for estimating evaporative potential (im/clo) from ASTM standard single wind velocity measures. In: Technical Report, T16-14, ADA#637325. Natick, MA, 01760, USA: US Army Research Institute of Environmental Medicine; 2016. Available from: www.dtic.mil/dtic/tr/fulltext/u2/a637325.pdf

[93] Kraning KK II, Gonzalez RR. A mechanistic computer simulation of human work in heat that accounts for physical and physiological effects of clothing, aerobic fitness, and progressive dehydration. Journal of Thermal Biology. 1997;**22**(4):331-342

[94] Welles AP, Tharion WJ, Potter AW, Buller MJ. Novel method of estimating metabolic rates of soldiers engaged in chemical biological defense training. In: Technical Report, T17-02, ADA#1022691. Natick, MA, 01760, USA: US Army Research Institute of Environmental Medicine; 2017

[95] McCullough EA, BW J, Huck J. A data base for estimating clothing insulation. ASHRAE Transactions. 1985;**91**:29-47

[96] Al-ajmi FF, Loveday DL, Bedwell KH, Havenith G. Thermal insulation and clothing area factors of typical Arabian Gulf clothing ensembles for males and females: Measurements using thermal manikins. Applied Ergonomics. 2008;**39**:407-414

[97] ISO 9920:2007. Ergonomics of the thermal environment. Estimation of thermal and insulation and evaporative resistance of a clothing ensemble. Geneva: International Organisation for Standardisation; 2007

[98] Osczevski R, Bluestein M. The new wind chill equivalent temperature chart. Bulletin of the American Meteorological Society. 2005;**86**(10):1453-1458

[99] National Weather Service. Windchill Temperature Index. Office of

Climate, Water, and Weather Services. Washington, D.C.: National Oceanic and Atmospheric Administration; 2001

[100] Xu X, Tikuisis P. Thermoregulatory modeling for cold stress. Comprehensive Physiology. 2014;**4**:1-25

[101] Gonzalez RR, McLellan TM, Withey WR, Chang SK, KB P. Heat strain models applicable for protective clothing systems: Comparison of core temperature response. Journal of Applied Physiology. 1997;**83**(3):1017-1032

[102] Department of the Army and Air Force. In: Heat Stress Control and Heat Casualty Management. TB-MED 507. Washington, DC: Government Printing Office; 2003

[103] Institute of Medicine. Dietary Reference Intakes for Water, Potassium, Sodium, Chloride, and Sulfate. Washington, DC: The National Academies Press; 2005

[104] Holmér I. Required clothing insulation (IREQ) as an analytical index of cold stress. ASHRAE Transactions. 1984;**90**(6):1116-1128

[105] ISO 11079. Ergonomics of the thermal environment—determination and interpretation of cold stress when using required clothing insulation (IREQ) and local cooling effects. Geneva: International Organisation for Standardisation; 2007

[106] Potter AW, Gonzalez JA, Carter AJ, Looney DP, Rioux TP, Srinivasan S, et al. Comparison of Cold Weather Clothing Biophysical Properties: US Army, Canadian Department of National Defence, and Norwegian Military. United States: US Army Research Institute of Environmental Medicine Natick; 2018

[107] Charkoudian N, Johnson JM. Reflex control of cutaneous vasoconstrictor system is reset by exogenous female reproductive hormones. Journal of Applied Physiology. 1999;**87**:381-385

[108] Charkoudian N, Stephens DP, Pirkle KC, Kosiba WA, Johnson JM. Influence of female reproductive hormones on local thermal control of skin blood flow. Journal of Applied Physiology. 1999;**87**:1719-1723

[109] Hodges GJ, Kosiba WA, Zhao K, Alvarez GE, Johnson JM. The role of baseline in the cutaneous vasoconstrictor responses during combined local and whole body cooling in humans. American Journal of Physiology Heart & Circulatory Physiology. 2007;**293**:H3187-H3192

[110] Johnson JM, Yen TC, Zhao K, Kosiba WA. Sympathetic, sensory, and nonneuronal contributions to the cutaneous vasoconstrictor response to local cooling. American Journal of Physiology Heart & Circulatory Physiology. 2005;**288**:H1573-H1579

[111] Stephens DP, Aoki K, Kosiba WA, Johnson JM. Nonnoradrenergic mechanism of reflex cutaneous vasoconstriction in men. American Journal of Physiology Heart & Circulatory Physiology. 2001;**280**:H1496-H1504

[112] Thompson CS, Holowatz LA, Kenney WL. Attenuated noradrenergic sensitivity during local cooling in aged human skin. Journal of Physiology. 2005;**564**:313-319

[113] Thompson CS, Kenney WL. Altered neurotransmitter control of reflex vasoconstriction in aged human skin. Journal of Physiology. 2004;**558**:697-704

[114] Thompson-Torgerson CS, Holowatz LA, Flavahan NA, Kenney WL. Rho kinase-mediated local cold-induced cutaneous vasoconstriction is augmented in aged human skin. American Journal of Physiology Heart & Circulatory Physiology. 2007;**293**:H30-H36

[115] Wissler EH. A quantitative assessment of skin blood flow in humans. European Journal of Applied Physiology. 2008;**104**:145-157

[116] Cooper KE, Edholm OG, Mottram RF. The blood flow in skin and muscle of the human forearm. The Journal of Physiology. 1955;**128**(2):258-267

[117] Nadel ER, Horvath SM, Dawson CA, Tucker A. Sensitivity to central and peripheral thermal stimulation in man. Journal of Applied Physiology. 1970;**29**:603-609

[118] Timbal J, Boutelier C, Loncle M, Bougues L. Comparison of shivering in man exposed to cold in water and in air. Pflügers Archiv. 1976;**365**:243-248

[119] Hayward JS, Eckerson JD, Collis ML. Thermoregulatory heat production in man: Prediction equation based on skin and core temperatures. Journal of Applied Physiology. 1977;**42**:377-384

[120] Tikuisis P, Giesbrecht GG. Prediction of shivering heat production from core and mean skin temperatures. European Journal of Applied Physiology and Occupational Physiology. 1999;**79**:221-229

[121] Parsons R. ASHRAE Handbook—Fundamentals. Atlanta, GA: American Society of Heating, Refrigerating and Air-conditioning Engineers; 1997

[122] Ainsworth BE, Haskell WL, Whitt MC, Irwin ML, Swartz AM, Strath SJ, et al. Compendium of physical activities: An update of activity codes and MET intensities. Medicine & Science in Sports & Exercise. 2000;**32**(suppl. 1):S498-S504

[123] Richmond PW, Potter AW, Looney DP, Santee WR. Terrain coefficients for predicting energy costs of walking over snow. Applied Ergonomics. 2019;**74**:48-54

[124] Looney DP, Potter AW, Pryor JL, Bremmer PE, Chalmers CR, McCLung HL, et al. Metabolic costs of standing and walking in healthy adults: A meta-egression. Medicine & Science in Sports & Exercise. 2019;**51**(2)

[125] American College of Sports Medicine. ACSM's Guidelines for Exercise Testing and Prescription. Philadelphia, PA: Lippincott Williams & Wilkins; 2013

[126] Balogun JA, Martin DA, Clendenin MA. Human energy expenditure during level walking on a treadmill at speeds of 54-130 m min-1. International Disability Studies. 1989;**11**(2):71-74

[127] Browning RC, Reynolds MM, Board WJ, Walters KA, Reiser RF. Obesity does not impair walking economy across a range of speeds and grades. Journal of Applied Physiology. 2013;**114**(9):1125-1131

[128] Pandolf KB, Givoni B, Goldman RF. Predicting energy expenditure with loads while standing or walking very slowly. Journal of Applied Physiology. 1977;**43**(4):577-581

[129] Workman JM, Armstrong BW. Metabolic cost of walking: Equation and model. Journal of Applied Physiology. 1986;**61**(4):1369-1374

[130] Santee WR. Windchill index and military applications. Aviation, Space, and Environmental Medicine. 2002;**73**:699-702

[131] Santee WR, Reardon MJ, Pandolf KB. Modeling the physiological and medical effects of exposure to environmental extremes. In: Friedl KE, Santee WR, editors. Military Quantitative Physiology: Problems and Concepts in Military Operational Medicine. Fort Detrick, MD: Office of The Surgeon General United States Army; 2012. pp. 39-72

[132] Castellani JW, O'Brien C, Baker-Fulco C, Sawka MN, Young AJ. Sustaining Health & Performance in Cold Weather Operations. USARIEM TN/02-2, ADA 395745. Natick, MA: US Army Research Institute of Environmental Medicine; 2001

[133] Friedl KE. Predicting human limits—The special relationship between physiology research and the Army mission. In: Friedl KE, Santee WR, editors. Military Quantitative Physiology: Problems and Concepts in Military Operational Medicine: Problems and Concepts in Military Operational Medicine. Fort Detrick, MD: Office of The Surgeon General United States Army; 2012. pp. 1-38

[134] den Hartog EA, Lotens WA. Postmortem time estimation using body temperature and a finite-element computer model. European Journal of Applied Physiology. 2004;**92**:734-737

[135] Mall G, Eisenmenger W. Estimation of time since death by heat-flow finite-element model. Part I: Method, model, calibration and validation. Legal Medicine (Tokyo, Japan). 2005;7:1-14

[136] Mall G, Eisenmenger W. Estimation of time since death by heat-flow finite-element model part II: Application to non-standard cooling conditions and preliminary results in practical casework. Legal Medicine (Tokyo, Japan). 2005;7:69-80

[137] Kanawaku Y, Kanetake J, Komiya A, Maruyama S, Funayama M. Computer simulation for postmortem cooling processes in the outer ear. Legal Medicine. 2007;**9**:55-62

[138] Smart JL. Estimation of time of death with a fourier series unsteady state heat transfer model. Journal of Forensic Science. 2010;**55**:1481-1487

[139] Smart JL, Kaliszan M. The post mortem temperature plateau and its role in the estimation of time of death. A review. Legal Medicine. 2012;**14**:55-62

[140] Smart JL, Kaliszan M. Use of a finite element model of heat transport in the human eye to predict time of death. Journal of Forensic Science. 2013;**58**:S69-S77

[141] Besnard Y, Launay JC, Guinet-Lebreton A, Savourey G. PREDICTOL: A computer program to determine the thermophysiological duration limited exposures in various climatic conditions. Computer Methods and Programs in Biomedicine. 2004;**76**:221-228

[142] Xu X, Amin M, Santee WR. Probability of Survival Decision Aid (PSDA). USARIEM T08/05, ADA478415. Natick, MA: US Army Research Institute of Environmental Medicine; 2008

[143] Keefe AA, Tikuisis P. A Guide to Making Stochastic and Single Point Predictions Using the Cold Exposure Survival Model (CESM). DRDC-TORONTO-TM-2008-061. Defence Research and Development Toronto (Canada); 2008

[144] Xu X, Allen A, Rioux T, Patel T, Sinha P, et al. Refinement of probability of survival decision aid (PSDA). In: Technical Note, TN14-02, ADA#599590. Natick, MA 01760 USA: U.S. Army Research Institute of Environmental Medicine; 2014

Chapter 5

Techniques to Reduce the Magnitude and Duration of Redistribution Hypothermia in Adults

Jonathan V. Roth

Abstract

While much effort has been devoted to correcting intraoperative hypothermia and documenting the adverse outcomes associated with hypothermia, less attention has been directed to preventing redistribution hypothermia in the first place. Methods currently exist that can reduce the magnitude of redistribution hypothermia, but are not widely practiced. This chapter focuses on the pathophysiology of redistribution hypothermia and the currently available methods that can be employed to reduce redistribution hypothermia. Additional promising, but currently unproven, methods are discussed. Since hypothermia causes adverse outcomes, it is anticipated that the reduction in redistribution hypothermia will improve patient outcome.

Keywords: redistribution hypothermia, hypothermia, perioperative hypothermia, intraoperative hypothermia, inhalation induction, anesthesia induction

1. Background

Hypothermia has multiple adverse consequences and should be avoided (**Table 1**) [1, 2]. The Anesthesia Patient Safety Foundation has recently reaffirmed that even mild hypothermia is associated with an increase in complications [3]. In studies assessing whether patients were hypothermic, typically the end-of-case temperature has been used for this determination and its association with complications. With the exception of one study where there was increased blood loss at 36.5°C [4], an increase in complications occurs when the end-of-case temperature is <36.0°C. However, there is increasing recognition that intraoperative temperature matters. The American College of Surgeons consider intraoperative hypothermia to be a modifiable risk factor for surgical site infections; they recommend the maintenance of intraoperative normothermia and the use of prewarming [5]. The 2017 Centers for Disease Control and Prevention (CDC) guidelines recommend maintenance of perioperative normothermia [6].

While much effort has been devoted to documenting adverse outcomes and correcting intraoperative hypothermia, relatively little attention has been directed to preventing intraoperative hypothermia in the first place. "Despite Active Warming, Hypothermia Is Routine in the First Hour of Anesthesia" was written on the cover of the February 2015 issue of Anesthesiology. In a

IntechOpen

• Morbid cardiac events (ischemia, infarctions, arrhythmias, sympathetic activation)
• Surgical wound infection
• Coagulopathy, increased blood loss, increased transfusion requirements
• Patient discomfort, postoperative shivering
• More likely to require postoperative ventilation
• Adverse respiratory events in PACU
• Delayed wake-up
• Prolonged PACU stays
• Increased hospital length of stay
• Negative nitrogen balance
• Delayed wound healing
• Increased financial cost of care of hypothermia complications
• Failure to meet MACRA standard

Table 1.
Complications of hypothermia.

retrospective review, Sun et al. found 64% of 58,814 adult patients had a temperature measurement under 36.0°C after 45 min [7]. Some hypothermia complications occur intraoperatively (e.g., coagulopathy and increased transfusion requirements), some postoperatively (e.g., shivering and delayed emergence) and some likely both (e.g., infection risk). The contribution of intraoperative hypothermia to postoperative complications may often be unrecognized. For example, patients may have decreased immunologic defense against infection at the time of incision, that is, during the vulnerable period when infections can become established. It is plausible that, if redistribution hypothermia can be reduced, one may be able to reduce the intraoperative and postoperative complications associated with hypothermia, particularly in situations where patients are at increased risk of developing a greater degree of hypothermia or may have increased risk of hypothermia-associated complications (**Table 2**). End-of-case hypothermia implies intraoperative hypothermia. End-of-case normothermia does not imply intraoperative normothermia. A patient may have been hypothermic intraoperatively, having suffered the consequences of intraoperative hypothermia, achieving normothermia only at the end of the case.

The body contains three thermal zones: the core (abdomen, thorax, and brain), the periphery (the extremities), and the skin. At rest, the core temperature is 37.0°C (36.5–37.5°C) and the periphery is 2–4°C cooler. The skin temperature can approach ambient temperature. At rest, most of the basal heat production occurs in the core. Heat travels from the core to the periphery to the skin and out to the environment. In the steady state, the rate of heat loss equals the rate of heat production, and the heat content of the body remains the same. Since temperature is just a measurement that reflects heat content, the temperature remains the same. The body normally maintains core temperature within a narrow range. Within limits, the periphery can act as a temperature buffer as it can add or lose heat, changing its temperature, while keeping the core temperature within a narrow range. The core temperature is the temperature that is physiologically most important [8].

There are behavioral (e.g., seeking an environment of a different temperature and changing clothing) and physiologic defenses to thermal challenges. Under anesthesia only the physiologic defenses are available. As one becomes too warm, the first physiologic defense is to vasodilate. If the temperature increases further,

Risk posed by postoperative hyperdynamic/tachycardic response to hypothermia
Coronary artery disease
Stenotic valvular heart disease
Dynamic obstructive cardiomyopathies
Increased risk or consequence of infection
Immunocompromised
Colon surgery
Foreign body placement (e.g., artificial joints)
Potential for large blood loss increased by hypothermia- induced coagulopathy
Spine surgery
Liver surgery
Prostate resection
Large exposure of tissues that have a propensity to bleed
Hypercarbia exacerbating hypothermia-induced coagulopathy
Increased risk of hypothermia due to patient characteristics
Elderly
Frail
Inability or delay in warming patient or environment
Lateral or prone positioning
Other prolonged positioning
Robotic surgery
Axillary-bifemoral artery bypass
Large surface area burn
Remote location with inability to adjust ambient temperature
Warming devices not available
Risk from hypothermia- induced vasoconstriction
Vascular surgery
Raynaud's disease or syndrome
Free flap with arterial vascular anastomosis

Table 2.
Examples of situations where patients are at increased risk of developing a greater degree of hypothermia or may have increased risk of hypothermia-associated complications.

the patient perspires. As one cools, the first defense is to vasoconstrict. If the temperature decreases further, the patient shivers [9]. These physiologic defenses are impaired during anesthesia.

There is a large vascular supply to the periphery and skin, but at rest these vessels are relatively vasoconstricted and there is relatively little blood flow. The blood flow to the periphery and skin can increase if these blood vessels vasodilate because of the administration of a vasodilator (or there is an increased metabolic need such as what occurs during physical activity). If pharmacologic-induced vasodilation occurs, the increased blood flow to the periphery transfers more heat from the core to the periphery and skin. As a result, the core's temperature decreases while that of the periphery will increase. This process is called redistribution hypothermia. Since heat only travels from higher to lower temperature (second law of thermodynamics), the heat in the periphery cannot be transferred back to the core. However, warming the periphery decreases the temperature gradient between the core and periphery. A smaller temperature gradient reduces the rate of heat transfer from the core [10]. Thus, more of the heat produced in the core will remain in the core, thus contributing to increasing the core temperature or decreasing the rate of core temperature decrease. If the periphery is warmed to a temperature greater than the core, heat can be transferred from the periphery to the core.

Propofol administration causes vasodilation and thus redistribution hypothermia. Propofol inductions typically result in a decrease in core temperature of about 1.5°C [11–13]. While there is also heat loss to the environment (via conduction, convection, radiation, evaporation, and airway losses), redistribution hypothermia is the major reason for the core temperature decrease in the first 15–60 min of an anesthetic. Although not specified in Sun's results, because propofol is the most common method of anesthetic induction in developed nations, it is likely most of these patients were induced with intravenous propofol and can explain the 64% incidence hypothermia (core temperature < 36.0°C) found in his review [7].

With this understanding, the following physiologic strategies have been studied to reduce redistribution hypothermia: (1) reduce the increased blood flow to the periphery and skin, (2) prewarm the periphery and skin, (3) increase metabolic activity, and (4) warm the environment. This chapter will discuss actual and potential methods available to reduce the magnitude and duration of redistribution hypothermia in adults.

2. Studied methods to reduce redistribution hypothermia

2.1 Reducing the increased blood flow to the periphery and skin

2.1.1 Etomidate

Compared to propofol, etomidate inductions result in a lesser initial temperature drop (1.4 vs. 0.5°C) [12]. Because of the adrenal axis suppression resulting from etomidate [14], the author does not recommend using etomidate just for thermal stability. However, if etomidate is used for other indications, one would expect a thermal benefit.

2.1.2 Ketamine

Compared to propofol, ketamine inductions result in a lesser initial temperature drop (1.5 vs. 0.9°C) [13]. Because of the risk of emergence reactions and hallucinations from an anesthetic dose of ketamine [15], the author does not recommend using ketamine just for thermal stability. However, if an anesthetic dose of ketamine is used for other indications, one would expect a thermal benefit.

2.1.3 Phenylephrine infusion

Ikeda et al. have demonstrated that a phenylephrine infusion of 0.5 mcg/kg/min starting immediately before induction with 2.5 mg/kg propofol results in an initial lower temperature decrease compared to propofol after the first hour (1.2 vs. 0.5°C decrease after 1 h) [16]. Presumably the vasoconstriction from phenylephrine opposes the vasodilation resulting from propofol administration. In addition, the patients who received the phenylephrine infusion maintained a higher mean arterial blood pressure (83 ± 9 vs. 72 ± 8 mm Hg, mean ± SD). (It seems plausible that any technique discussed in this section that reduces vasodilation has the potential to accrue an additional benefit of reducing induction-associated hypotension. This hypothesis requires investigation.)

2.1.4 Phenylephrine bolus

A 160 mcg bolus of phenylephrine immediately prior to 2.2 mg/kg propofol reduces the mean decrease in core temperature by about 0.43°C in the first hour than those who did not receive the phenylephrine bolus [17, 18]. While redistribution hypothermia can continue for up to 3 h, a large part of the temperature decrease occurs within the first 15 min. The vasoconstricting effect of a bolus of phenylephrine lasts sufficiently long to oppose much of the maximal vasodilation resulting from propofol induction. While most patients decrease their blood pressure after propofol administration, the bolus phenylephrine reduced the incidence of propofol-induced hypotension from 98 to 58% [17, 18]. While generally effective, the 160 mcg dose was used on all patients in this study but may not be optimal. Some patients still became hypotensive (systolic BP < 85 mm Hg), and 1 patient in this group of 50 patients increased the systolic blood pressure to >180 mm Hg [17, 18]. It remains to be determined if a weight-based dose could be found that further reduces the incidence of hypotension, avoids dangerous hypertension, and still maintains the thermal benefit.

2.1.5 Inhalation inductions

Ikeda et al. demonstrated less core hypothermia when anesthesia is induced with inhaled sevoflurane than with intravenous propofol (1.5 vs. 0.8 °C decrease after 1 h) [11]. This study of 10 patients in each group was done at a time when the concept of redistribution hypothermia was still in development and the harmful effects of even mild hypothermia were not as well appreciated as they are today. A recent study (50 patients in each of six groups) replicated and strengthened these findings [17, 18]. Inhalation inductions of 8% sevoflurane in either 100% oxygen or 50% oxygen/50% nitrous oxide resulted in a higher mean temperature by about 0.5°C than those who received 2.2 mg/kg propofol in patients aged 18–55 years [17, 18]. Inhalation inductions were also found effective in reducing redistribution hypothermia in older (56–88 years, mean 67.2 years) patients. Elderly patients have an increased risk for hypothermia [19–21] for reasons that include decreased metabolic activity, decreased muscle mass, an impaired vasoconstriction response, and an impaired shivering response. A previous study also concluded that inhalation induction is more hemodynamically stable than IV propofol inductions [22]. In contrast to propofol inductions where significant hypotension can occur immediately, an inhalation induction typically causes a more gradual decrease in blood pressure which can be treated before severe hypotension develops.

In adults, anesthetic inductions are achieved most commonly by intravenous, not inhalation, inductions for reasons that include inhalation inductions take extra time, room contamination with anesthetic gases, and possible patient dissatisfaction. An inhalation induction takes 1–2 min longer than an intravenous induction [17, 18] and that lost time may be recovered by a quicker wake-up because of the patient being warmer. However, Muzi et al. demonstrated that the speed of inhalation induction approached that of an intravenous induction using a primed circuit [23]. Although many anesthesia practitioners may assume patients would not want the inhalation technique, when offered a choice, 50% chose an inhalation induction, 33% chose IV induction, and 17% were undecided [24].

Inhalation inductions are not for everyone. Medical contraindications would include concern of increased intracranial pressure, indication for hypothermia, contraindication to hyperthermia (e.g., multiple sclerosis), increased aspiration risk, unfavorable airway anatomy, and patient fear of face masks. Since patients may lighten more rapidly when the face mask is removed for endotracheal intubation than with propofol, it may be prudent to avoid inhalation inductions when intubation may be a more prolonged process as there may potentially be an increased risk of awareness than a propofol induction. Examples would include inserting double-lumen tubes or training novice laryngoscopists.

However, there are additional potential benefits to preforming inhalation inductions. First, there will be no pain on propofol injection. Second, trainees will get more practice with airway management. In current practice, most patients after IV induction immediately receive a laryngeal mask airway (LMA) or endotracheal intubation. Third, future propofol shortages can be mitigated by employing inhalation inductions. Fourth, LMAs may be easier to insert while patients are breathing spontaneously as the airway tends to open during inspiration and there is less of an obstruction to proper LMA positioning than a totally collapsed airway one typically gets after IV propofol inductions. Fifth, there will be less second-hand exposure to propofol, currently a candidate factor in propofol addiction. Sixth, inhalation inductions may be a superior alternative over other induction agents to patients with allergies to propofol. Seventh, with intravenous inductions, atelectasis develops very quickly. One would expect that with spontaneous ventilation, there may be less atelectasis, but this will need to be studied. In patients breathing spontaneously via an LMA after IV propofol induction, one does not have to manage a patient who becomes apneic, thus eliminating extra tasks and saving time while starting a case. Lastly, propofol supports bacterial growth [25]. There is an increased number of colony-forming units in the stopcocks of patients who received propofol (10× at 24 h and >100× at 48 h) compared to those who did not [26]. While it is not established that this is a cause of increased infections, the avoidance of propofol would eliminate this as a concern. Removing the stopcocks could also address this concern but that adds cost and likely would not be universally done.

2.1.6 Nitrous oxide

Previous work suggests an ongoing thermal benefit to using nitrous oxide. Ozaki et al. found nitrous oxide impairs thermoregulation less than sevoflurane or isoflurane [27]. The threshold for vasoconstriction was 35.8 ± 0.3°C (mean ± SD) in the patients given 50% nitrous oxide combined with 0.5 MAC sevoflurane, which was statistically significantly greater than that in those given 1.0 MAC sevoflurane: 35.1 ± 0.4°C. Similarly, the threshold for vasoconstriction was 35.9 ± 0.3°C in the patients given 60% nitrous oxide combined with 0.5 MAC isoflurane, which was statistically significantly greater than that in those given 1.0 MAC isoflurane: 35.0 ± 0.5°C. The use of nitrous oxide allows for the thermal defense of vasoconstriction to activate before the patient becomes more hypothermic.

Nitrous oxide has been under challenge for several decades. Two of the reasons why nitrous oxide has been out of favor with many practitioners have been the concern of major cardiovascular morbidity and mortality and an increased risk of surgical site infections (SSI). In combination with another retrospective study of 49,016 patients where nitrous oxide use was associated with decreased 30-day mortality and decreased in-hospital mortality/morbidity, the results of the ENIGMA II have essentially eliminated these concerns [28–31]. ENIGMA II concludes "Our findings support the safety profile of nitrous oxide use in major non-cardiac surgery. Nitrous oxide did not increase the risk of death and cardiovascular complications or surgical site infection, the emetogenic effect of nitrous oxide can be controlled with antiemetic prophylaxis, and a desired effect of reduced volatile agent use was shown." [4] The other major reason for not using nitrous oxide has been the concern for postoperative nausea and vomiting (PONV). If a patient has been administered an antiemetic, there is a small nonsignificant increased risk of severe PONV. ENIGMA II concludes "Nitrous oxide increases the risk of severe PONV by only a small percentage, and the increased risk is essentially eliminated by antiemetic drug prophylaxis. Concern about severe PONV thus does not appear to be a valid reason to avoid nitrous oxide."

Except for potential environmental concerns, there is little reason not to use nitrous oxide in cases that are not of long duration (>4–6 h) unless there are physical contraindications (e.g., gas space expansion). Besides from its potential thermal benefit, nitrous oxide has been shown to reduce chronic pain in specific populations (Asians and other patients with variants in the methylenetetrahydrofolate reductase gene) [32]. The United States is in the midst of an opioid epidemic. The majority of heroin users got their start from medically prescribed opioids [33]. Nitrous oxide also has analgesic efficacy and may reduce intraoperative opioid use. Further research is needed, but the possibility of reducing chronic pain and intraoperative opioid use may have benefit in combatting the opioid epidemic [34].

2.2 Prewarm the periphery and skin

Prewarming is the active warming of the body surface, often via forced-air warming, prior to induction of general or central neuraxial anesthesia. It is currently the most effective method of reducing redistribution hypothermia. It has been extensively studied, and, in addition to demonstrating warmer core temperatures, improved outcomes (decreased blood loss, transfusion requirement, and infection rate) have been demonstrated. (A recent chapter reviews much of the relevant detail and will not be repeated here [10]. A small representative sample of studies are listed [35–39].) Prewarming is fundamentally different from all other techniques in that it's the only technique that exogenously adds heat content to the patient. However, the technique is not universally used [40]. Obstacles to its use include (1) requirement of space, equipment, supplies, and personnel time, (2) change in the pattern of patient flow, (3) patient refusal or intolerance, (4) requirement of cleaning if reusable equipment is utilized, (5) insufficient availability of a power supply, (6) requirement to train personnel, (7) bypass of the holding area, (8) additional equipment maintenance requirement, and (9) inadequate knowledge of the value of prewarming [10].

Prewarming works by adding heat content to the periphery. This decreases the temperature gradient between core and periphery and thus decreases the heat transfer and redistribution hypothermia. Any method that can increase the peripheral temperature will reduce redistribution hypothermia. Any event that decreases peripheral temperature will increase redistribution. Thus, all reasonable efforts should be made to keep the periphery warm before induction of anesthesia. After application of forced-air warming, it will take time (usually 30 min) until an increase in core temperature occurs [41, 42]. This delay occurs because the periphery needs to be warmed before there is a significant effect on core temperature.

The efficacy of prewarming can be limited by sweating, thermal discomfort, and efficacy of the warming device. Sessler et al. found that 30 min of prewarming increased peripheral tissue heat content by more than the amount normally distributed during the first hour of anesthesia [43]. Since there are other and ongoing mechanisms of heat loss, prewarming more than 30 min will likely benefit many patients. However, if it is difficult to arrange for 30+ min of prewarming or the patient does not tolerate the longer durations, even 10–20 min of prewarming is effective in reducing hypothermia and shivering [44].

Because of redistribution hypothermia, ideally, every patient undergoing general or neuraxial anesthesia should be prewarmed [45, 46]. If the patient receives just a peripheral nerve block, there is little risk of hypothermia. Prewarming (and forced-air warming) should not be applied over ischemic limbs. Normally when there is heat transfer to an area of the body, blood circulation removes the heat from that area, thus decreasing the local temperature. If there is impaired blood flow, it is possible that the heat accumulation from prewarming or intraoperative forced-air warming could cause tissue damage. (In therapeutic hyperthermia, temperatures >42.0°C have been associated with tissue damage such as fat necrosis [47].) For similar reasons, forced-air warming over the lower extremities should be turned off during aortic cross-clamping. Also, in theory, there may be more risk of cell death from warming ischemic tissue because of the resulting increase in metabolic oxygen demand in combination with the impaired blood supply. It may be prudent to avoid prewarming when there is a contraindication to hyperthermia (e.g., risk of neurologic ischemia and pregnancy).

There is no data to guide the decision to use prewarming on patients who are hyperthermic preoperatively. Patients are hyperthermic because either (1) their cooling mechanisms have been overwhelmed as that which occurs in heat exhaustion or heatstroke (nonfebrile hyperthermia) or (2) they have an elevated temperature set point as occurs with many infections (febrile hyperthermia). The nonfebrile patients probably should be allowed to have their core temperature normalized and thus probably should not be prewarmed. It has been suggested that the febrile patients should be allowed to remain hyperthermic intraoperatively [48]. There is overwhelming evidence that fever is part of a coordinated defense system [49, 50]. The lines of evidence include evolutionary, correlative, antipyretic, and hyperthermia/hypothermia studies [49]. For example, infectious illnesses in animals are of longer duration, and mortality rates increase if the fever is treated [49]. Some of the enzymes in the immune system have a temperature optima in the febrile range. In addition, if the temperature of these patients decrease to below their elevated temperature set point and the set point does not change during the anesthetic, then these patients will behave postoperatively as though they are hypothermic (e.g., increasing metabolism and cardiac output, shivering), even if their temperature is >37.0°C. Thus, although unproven, there is reason to maintain the febrile hyperthermia intraoperatively. It is an unanswered question as to whether these patients should be prewarmed.

2.3 Increase metabolic activity

2.3.1 Amino acid administration

The preoperative administration of amino acids increases metabolic heat production and leads to the release of insulin and leptin resulting in a mean temperature increase of 0.46°C [51]. These hormones may also affect central thermoregulation. If amino acid infusion is started after hypothermia develops, rewarming is not augmented [52]. It is possible that the amino acid-induced increase in cardiopulmonary demands may be problematic in frail patients and those with reduced cardiopulmonary reserve. Since there is limited evidence, this technique is considered experimental.

2.3.2 Fructose administration

The preoperative administration of fructose increases metabolic heat production and affects central thermoregulation [53]. However, in patients with hereditary fructose intolerance (HFI), the infusion of fructose can lead to liver damage, kidney damage, convulsions, and death. HFI often goes undiagnosed. The prevalence of HFI is estimated at 1 in 20,000, similar to the incidence of malignant hyperthermia events.

2.4 Warming the environment

As discussed above, anything that is practical and can be done to keep the patient warmer will likely result in the periphery remaining warmer and thus less redistribution. There is often a difference of opinion among various operating room personnel as to what temperature of the operating room should be. A cooler environment will increase the rate of heat loss from the patient. With the resultant decrease in peripheral heat content, the magnitude of redistribution hypothermia will be greater [8].

There are five methods of heat loss (conduction, convection, radiation, evaporation, and loss via the airway). Radiation and convection losses are most important [54]. One of the major determinants of radiative heat loss is the temperature difference between the radiator (i.e., the patient) and the environment. A greater temperature difference will result in a greater heat loss. Another major determinant is the absorption/reflection properties of the environment. The author is unaware of any clinical data regarding these factors.

Convection refers to heat transfer resulting from the bulk movement of a fluid (i.e., gas or liquid). A patient will transfer heat to warm the air immediately around him or herself. Convective airflow will move this warm air away from the patient and replace it with cooler ambient air. Thus, heat loss will continue to warm the newly adjacent cool air. The cooler the adjacent air, the greater the rate of heat loss from the patient.

Surgeons generally prefer a cooler room because they are working, are under lights that may emit heat, may be under stress, are gowned, may be in physical contact with other personnel, and may also be wearing lead aprons. An uncomfortable surgeon may not work at his/her best and may drip perspiration into the surgical wound. With modern operating rooms where the air is replaced many times an hour, the temperature can be adjusted within minutes. Thus, a reasonable compromise would be to keep the operating room warm until the patient is prepped and draped and then cool the room for the benefit of the surgical team. Once the patient is draped, convective losses are reduced except from the surgical wound.

3. Candidate methods to reduce redistribution hypothermia

Unfortunately, none of the abovementioned techniques fully solves the redistribution hypothermia problem. It is plausible that either reducing propofol dosages or combining techniques may provide additional thermal benefit. The following techniques show promise but require formal investigation:

1. Ketamine in analgesic doses is commonly used as part of a multimodal analgesia strategy. It is plausible that reducing the propofol dose by an analgesic dose of ketamine would reduce the magnitude of redistribution hypothermia. The induction dose of propofol (2.2 mg/kg) is similar in mg to the induction dose of ketamine (2 mg/kg). Reducing the propofol dose by 30 mg and replacing it with 30 mg ketamine seems reasonable.

2. Kazama et al. found that patients can be induced with a reduced total dose of propofol and with less hypotension when diluted propofol was administered

as an infusion [55]. It is plausible that, by using less propofol, there would be a lesser amount of redistribution hypothermia (and less hypotension).

3. A blended propofol-inhalation induction would utilize less propofol and thus potentially reduce redistribution hypothermia.

4. Combining prewarming with any of the other techniques (e.g., prewarming and inhalation induction, prewarming and phenylephrine prior to propofol).

5. Combining prophylactic phenylephrine with inhalation inductions.

4. Summary

At this time, prewarming is the most studied and likely the most effective method of reducing redistribution hypothermia, and improved outcomes have been documented. Unfortunately, it is not universally used. Given the priority of operating room expediency, either inhalation inductions or prophylactic administration of bolus phenylephrine are practical and can be used in virtually every anesthetizing location. Even though these techniques have been demonstrated to reduce redistribution hypothermia, and post-induction temperatures are similar to what one sees after prewarming and a propofol induction, we can only anticipate but not yet infer the same improved outcomes will accrue. Although a strong correlation of adverse outcomes and hypothermia has been documented in numerous studies, an outcome study is needed. Inhalation inductions or prophylactic administration of phenylephrine reduces redistribution hypothermia by reducing vasoconstriction; they do not add heat content. Prewarming reduces redistribution hypothermia by warming the periphery and adds heat content to the patient. Because the periphery needs to get warmed before forced-air warming increases the core temperature, it is likely that prewarmed patients will rewarm more rapidly, which is likely beneficial.

It is important to keep the operating room warm until the patient is prepped and draped. The temperature of a modern operating room can be decreased rapidly for the comfort of the operating room personnel. Putting a warm blanket on a patient as he/she enters a cold operating room does little to rewarm a patient. The skin temperature receptors have a disproportionate influence on the hypothalamus. The warm blanket may make the patient feel warmer, but the patient may still have lost significant heat content to the cool environment.

Besides from thermal benefits, financial benefits may accrue from reducing redistribution hypothermia. In the United States, the new Medicare Access and CHIP Reauthorization Act of 2015 (MACRA) temperature target (35.5°C) may now be easier to achieve [56]. Avoidance of unpleasant side effects (e.g., shivering) may result in less patient dissatisfaction. Reducing hypothermia-associated complications will reduce costs.

Author details

Jonathan V. Roth[1,2*]

1 Albert Einstein Medical Center, Philadelphia, PA, USA

2 Sidney Kimmel Medical College - Thomas Jefferson University, Philadelphia, PA, USA

*Address all correspondence to: jvroth1@aol.com

IntechOpen

© 2020 The Author(s). Licensee IntechOpen. This chapter is distributed under the terms of the Creative Commons Attribution License (http://creativecommons.org/licenses/by/3.0), which permits unrestricted use, distribution, and reproduction in any medium, provided the original work is properly cited.

References

[1] Sessler DI. Complications and treatment of mild hypothermia. Anesthesiology. 2001;**95**:531-543

[2] Stewart PA, Liang SS, Li QS, Huang ML, Bilgin AB, Kim D, et al. The impact of residual neuromuscular blockade, oversedation, and hypothermia on adverse respiratory events in a postanesthetic care unit: A prospective study of prevalence, predictors, and outcomes. Anesthesia and Analgesia. 2016;**123**(4):859-868

[3] Prielipp RC, Birnbach DJ. HCA-infections: Can the anesthesia provider be at fault? APSF Newsletter. 2018;**32**(2):64-65

[4] Winkler M, Akca O, Birkenberg B, Hetz H, Scheck T, Arkilic CF, et al. Aggressive warming reduces blood loss during hip arthroplasty. Anesthesia and Analgesia. 2000;**91**:978-984

[5] Ban KA, Minei JP, Laronga C, Harbrecht BG, Jensen EH, Fry DE, et al. American College of Surgeons and Surgical Infection Society: Surgical Infection Guidelines, 2016 update. Journal of the American College of Surgeons. 2017;**224**:59-74

[6] Berríos-Torres SI, Umscheid CA, Bratzler DW, Leas B, Stone EC, Kelz RR, et al. Centers for disease control and prevention guideline for the prevention of surgical site infection, 2017. JAMA Surgery. 2017;**152**(8):784-791

[7] Sun Z, Honar H, Sessler DI, Dalton JE, Yang D, Panjasawatwong K, et al. Intraoperative core temperature patterns, transfusion requirement, and hospital duration in patients warmed with forced air. Anesthesiology. 2015;**122**:276-285

[8] Brauer A. Influence of transportation to the operating room and preparation for surgery. In: Perioperative Temperature Management. United Kingdom: Cambridge University Press; 2017. pp. 42-43. Chapter 6

[9] Brauer A. Physiology of thermoregulation. In: Perioperative Temperature Management. United Kingdom: Cambridge University Press; 2017. pp. 17-25. Chapter 3

[10] Brauer A. Prewarming. In: Perioperative Temperature Management. United Kingdom: Cambridge University Press; 2017. pp. 170-177. Chapter 32

[11] Ikeda T, Sessler DI, Kikura M, Kazama T, Ikeda K, Sato S. Less core hypothermia when anesthesia is induced with inhaled sevoflurane than with intravenous propofol. Anesthesia and Analgesia. 1999;**88**:921-924

[12] Park HP, Kang JM, Jeon YT, Choi IY, Oh YS, Hwang JW. Comparison of the effects of etomidate and propofol on redistribution hypothermia during general anesthesia. Korean Journal of Anesthesiology. 2006;**50**:S19-S24

[13] Ikeda T, Kazama T, Sessler DI, Toriyama S, Niwa K, Shimada C, et al. Induction of anesthesia with ketamine reduces the magnitude of redistribution hypothermia. Anesthesia and Analgesia. 2001;**93**:934-938

[14] Lundy JB, Slane ML, Frizzi JD. Acute adrenal insufficiency after a single dose of etomidate. Journal of Intensive Care Medicine. 2007;**22**(2):111-117

[15] Garfeld JM, Garfield FB, Stone JG, Hopkins D, Johns LA. A comparison of psychologic responses to ketamine and thiopental-nitrous-halothane anesthesia. Anesthesiology. 1972;**36**:329-338

[16] Ikeda T, Ozaki M, Sessler DI, Kazama T, Ikeda K, Sato S. Intraoperative phenylephrine infusion decreases the magnitude of

redistribution hypothermia. Anesthesia and Analgesia. 1999;**89**(2): 462-465

[17] Roth JV, Braitman LE. Induction techniques that can reduce redistribution hypothermia. Abstract Presented at the American Society of Anesthesiologists Annual Meeting; October 23, 2016

[18] Roth JV, Braitman LE. Induction techniques that can reduce redistribution hypothermia. Abstract Presented at the International Anesthesia Research Society Annual Meeting, #1135; May 2017

[19] Vaughan MS, Vaughan RW, Cook RC. Postoperative hypothermia in adults: Relationship of age, anesthesia, and shivering to rewarming. Anesthesia and Analgesia. 1981;**60**:746-751

[20] Kurz A, Plattner O, Sessler DI, Huemer G, Redi G, Lackner F. The threshold for thermoregulatory vasoconstriction during nitrous oxide/ isoflurane anesthesia is lower in elderly than in young patients. Anesthesiology. 1993;**79**:465-469

[21] Sessler DI. Temperature monitoring and perioperative thermoregulation. Anesthesiology. 2008;**109**(2):318-338

[22] Thwaites A, Edmends S, Smith I. Inhalation induction with sevoflurane: A double-blind comparison with propofol. British Journal of Anaesthesia. 1997;**78**(4):356-361

[23] Muzi M, Robinson BJ, Ebert TJ. Induction of anesthesia and tracheal intubation with sevoflurane in adults. Anesthesiology. 1996;**85**:536-543

[24] Van den Berg AA, Chitty DA, Jones RD, Sohel MS, Shahen A. Intravenous or inhaled induction of anesthesia in adults? An audit of preoperative patient preferences. Anesthesia & Analgesia. 2005;**100**(5):1422-1424

[25] Thomas DV. Propofol supports bacterial growth. British Journal of Anaesthesia. 1991;**66**:274

[26] Cole DC, Baslanti TO, Gravenstein NL, Gravenstein N: Leaving more than your fingerprint on the intravenous line: A prospective study on propofol and implications of stopcock contamination. Anesthesia and Analgesia. 2015;**120**(4):861-867

[27] Ozaki M, Sessler DI, Suzuki H, Ozaki K, Tsunoda C, Atarashi K. Nitrous oxide decreases the threshold for vasoconstriction less than sevoflurane or isoflurance. Anesthesia and Analgesia. 1995;**80**(6):1212-1216

[28] Myles PS, Leslie K, Chan MTV, Forbes A, Peyton PJ, Pasch MJ, et al. ANZCA Trials Group for the ENIGMA-II Investigators: The safety of addition of nitrous oxide to general anaesthesia in at-risk patients having major non-cardiac surgery (ENIGMA-II): A randomized, single-blind trial. Lancet. 2014;**384**:1446-1454

[29] Turan A, Mascha EJ, You J, Kurz A, Shiba A, Saager L, et al. The association between nitrous oxide and postoperative mortality and morbidity after noncardiac surgery. Anesthesia and Analgesia. 2013;**116**:1026-1033

[30] Fleisher LA. Value of sequels. Is it safe to include nitrous oxide in your anesthetic. Anesthesiology. 2015;**123**(6):1229-1230

[31] Leslie K, Myles PS, Kasza J, Forbes A, Peyton PJ, Chan MTV, et al. Nitrous oxide and serious long-term morbidity and mortality in the evaluation of nitrous oxide in the gas mixture for anaesthesia (ENIGMA)-II trail. Anesthesiology. 2015;**123**(6):1267-1280

[32] Chan MT, Peyton PJ, Myles PS, et al. Chronic postsurgical pain in the evaluation of nitrous oxide in the gas

mixture for anaesthesia (EnNIGMA)-II trial. British Journal of Anaesthesia. 2016;**117**:801-811

[33] Quinones S. Dreamland: The True Tale of America's Opiate Epidemic. New York, NY: Bloomsbury Press; 2015

[34] Roth JV. Chronic pain and the opioid epidemic. Are we ignoring the potential benefits of nitrous oxide? Anesthesia & Analgesia. 2018;**126**(4):1423-1424

[35] Just B, Trevien V, Lelva E, Lienhart A. Prevention of intraoperative hypothermia by preoperative skin-surface warming. Anesthesiology. 1993;**79**:214-218

[36] Andrzejowski J, Hoyle J, eapen G, Turnbull D. Effect of prewarming on post-induction core temperature and the incidence of inadvertent perioperative hypothermia in patients undergoing general anaesthesia. British Journal of Anaesthesia. 2008;**101**:627-631

[37] Bock M, Muller J, Bach A, Bohrer H, Martin E, Motsch J. Effects of preinduction and intraoperative warming during major laparotomy. British Journal of Anaesthesia. 1998;**80**:159-163

[38] Camus Y, Delva E, Sessler DI, Lienhart A. Pre-induction skin-surface warming minimizes intraoperative core hypothermia. Journal of Clinical Anesthesia. 1995;7:384-388

[39] Hynson JM, Sessler DI, Moayeri A, McGuire J, Schroeder BS. The effects of pre-induction warming on temperature and blood pressure during propofol-nitrous oxide anesthesia. Anesthesiology. 1993;**79**:219-228

[40] Brauer A, Russo M, Nickel EA, Bauer M, Russo SG. Anwendungsrealitat des peripoperativen Warmemanagements in Deutschland. Ergebnisse einer Online-Umfrage. Anästhesiologie und Intensivmedizin. 2015;**56**:287-297

[41] Negishi C, Hasegawa K, Mukai S, Nakagawa F, Ozaki M, Sessler DI. Resistive heating and forced-air warming are comparatively effective. Anesthesia and Analgesia. 2003;**96**:1683-1687

[42] Kimberger O, Held C, Stadelmann K, Mayer N, Hunkeler C, Sessler DI, et al. Resistive polymer versus forced-air warming: Comparable heat transfer and core rewarming rates in volunteers. Anesthesia and Analgesia. 2008;**107**:1621-1626

[43] Sessler DI, Schroeder BA, Merrifield B, Matsukawa T, Cheng C. Optimal duration and temperature of prewarming. Anesthesiology. 1995;**82**:674-681

[44] Horn EP, Bein B, Bohm R, Steinfath M, Sahili N, Hocker J. The effect of short time periods of pre-operative warming in the prevention of perioperative hypothermia. Anaesthesia. 2012;**67**:612-617

[45] Matsukawa T, Sessler DI, Sessler AM, Schroeder BA, Ozaki M, Kurz A, et al. Heat flow and distribution during induction of general anesthesia. Anesthesiology. 1995;**82**:662-673

[46] Matsukawa T, Sessler DI, Christensen R, Ozaki M, Schroeder M. Heat flow and distribution during epidural anesthesia. Anesthesiology. 1995;**83**:961-967

[47] Mackowiak PA, Boulant JA. Fever's glass ceiling. Clinical Infectious Diseases. 1996;**22**:525-536

[48] Roth JV. Some unanswered questions about temperature management. Anesthesia & Analgesia. 2009;**109**(5):1695-1699

[49] Kluger MJ, Kozak W, Conn CA, Leon LR, Soszynski D. The adaptive value of fever. Infectious Disease Clinics of North America. 1996;**10**:1-20

[50] Mackowiak PA. Fever: Blessing or curse? A unifying hypothesis. Annals of Internal Medicine. 1994;**120**:1037-1040

[51] Aoki Y, Aoshima Y, Atsumi K, Kaminaka R, Nakau R, Yanagida K, et al. Perioperative amino acid infusion for preventing hypothermia and improving clinical outcomes during surgery unger general anesthesia: A systematic review and meta-analysis. Anesthesia and Analgesia. 2017;**125**:793-802

[52] Inoue S, Shinjo T, Kawaguchi M, Nakajima Y, Furuya H. Amino acid infusions started after development of intraoperative core hypothermia do not affect rewarming but reduce the incidence of postoperative shivering during major abdominal surgery: A randomized trial. Journal of Anesthesia. 2011;**25**:850-854

[53] Mizobe T, Nakajima Y, Ueno H, Sessler DI. Fructose administration increases intraoperative core temperature by augmenting both metabolic rate and the vasoconstriction threshold. Anesthesiology. 2006;**104**:1124-1130

[54] Brauer A. Physiology of Heat Gain and Heat Loss. United Kingdom: Cambridge University Press; 2017. pp. 26-32. Chapter 4

[55] Kazama T, Ikeda K, Morita K, Kikura M, Ikeda T, Kurita T, et al. Investigation of effective anesthesia induction doses using a wide range of infusion rates with undiluted and diluted propofol. Anesthesiology. 2000;**92**:1017-1028

[56] MIPS Standard #424. The Medicare Access and CHIP Reauthorization Act of 2015 (MACRA). Centers for Medicare and Medicaid Services. Federal Register; November 4, 2016

Section 3

Special Focus

Chapter 6

Heart Rate Variability as Biomarker for Prognostic of Metabolic Disease

Alondra Albarado-Ibañez, Rosa Elena Arroyo-Carmona, Daniela Alexandra Bernabé-Sánchez, Marissa Limón-Cantú, Benjamín López-Silva, Martha Lucía Ita-Amador and Julián Torres-Jácome

Abstract

Lifestyle emerging diseases like obesity, metabolic syndrome (MeS), and diabetes mellitus are considered high-risk factors for lethal arrhythmias and side effects. A Poincaré plot is constructed with the time series of RR and PP electrocardiogram (ECG) intervals, using two stages: the new phase and the old phase. We proposed this diagram of two dimensions, a way to quantify and observe the regularity of events in space and time. Therefore, the heart rate variability (HRV) can be used as a biomarker for early prognostic and diagnostic of several metabolic diseases; additionally, this biomarker is obtained by a noninvasive tool like the electrocardiogram.

Keywords: Poincaré plot, heart rate variability, metabolism, biomarker

1. Introduction

The biological phenomena could be explained by classical physics, and most of these phenomena are characterized by cycles. Usually, the time period required to "complete a cycle" is not constant. The study of time period fluctuations represents a way to assess interactions between other systems and the intrinsic properties of the same system.

The light/dark cycle (circadian rhythm) and the cyclic seasons that divide the year by changes on weather, ecology, and hour of daylight allowed the evolution of life on earth [1]. These series of events have influenced the organisms inducing cycles that are essential for life (hormonal, organ function, behavior, production of neurotransmitters, reproduction, and others); all cycles are fluctuations related to several biological phenomena. The study and knowledge of the fluctuations of biological phenomena are valuable to analyze the intrinsic properties of a system and the interaction with other cyclic systems [2].

The quantification of biological variability has been used to study several physiological phenomena, among them, fluctuations on the heart rate using the RR interval period of the electrocardiogram (ECG). The heart rate variability (HRV) is a useful health indicator [3], and in this chapter we detail how this tool is used for the prognosis and diagnosis of metabolic diseases.

IntechOpen

2. Biological variability

All organisms present dynamic and complex oscillations in their function. The time between every oscillation is called period, and it represents biological rhythms. These rhythms regulate all physiological processes with periods of milliseconds as neuronal activity, seconds as the heart rate, hours as hormone release, monthly as the ovarian cycle, and annually as the growth and migration. The biological rhythms are present in all levels of biological organization at the molecular, organelle, cell, and tissue levels; these organizations are present in vertebrates, invertebrates, and plants (**Figure 1**).

The study of period variations is essential because these fluctuations represent the interaction of the cycles with other systems or alterations on the intrinsic properties of the same system, individual and interspecific variability [4].

The periods can oscillate only under the influence of an external periodic signal originating exogenous rhythms; these allow changes in the variability of biological rhythm associated with external environmental synchronizer [5]. However, when the light/darkness external synchronizer is removed, a self-sustaining oscillation is shown, so it is said that the system has an autonomous endogenous rhythm.

The biological rhythms with a periodicity of 24 h are denominated circadian rhythms (*circa* = about, *diem* = a day). These circadian rhythms develop an endogen rhythm with one period close to 24 h under constant darkness, the free running, but can be synchronized again with the light and darkness; this phenomenon is called circadian entrainment. The circadian rhythms of longer period are infradians, such as the menstrual cycle, while shorter periods are ultradians, such as cardiac frequency, the autonomic system regulation, electrical activity of neurons, and secretion of hormones, among others (see **Figure 1**).

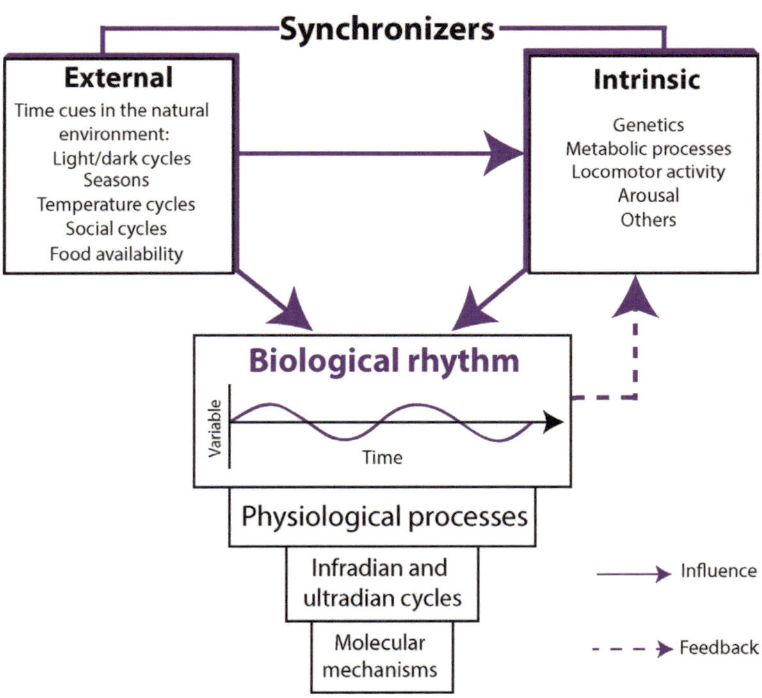

Figure 1.
Variability biologic system. All organisms develop the variability of biological systems for environmental adaptation. Physiological and metabolic processes depend on the interaction between the central and peripheral rhythms.

The autonomous nervous system (sympathetic and parasympathetic) regulates the cardiovascular system that involves the heart rate variability. The relevance that the intermittent oscillations of peripheral clocks modulate the variability of the central clock is fundamental for health process. The coordination and communication among peripheral and central clock are essential for metabolic, enzymatic, molecular, and physiological process.

3. Heart rate

The heart rate (HR) is determined by the activity of the sinoatrial node. The electrical activity propagates to the atria and then to the atrioventricular node, and, finally, the electrical activity reaches the ventricles triggering its contraction from apex to base. Any change in the origin and propagation of this electrical activity is denominated arrhythmia. The contraction and relaxation of cardiac tissue is a process named heartbeat. It is a cyclical event, the beats per minute produce the heart rate. Heart rate is a parameter that serves to diagnose some health problems in patients. When HR is increased, it is called tachycardia and the decrease of HR is called bradycardia. Tachycardia is related to exercise, emotions, the fight or flight phenomenon, among other activities. Bradycardia is related to sleep and rest. To measure HR there are several methods that are used in the clinic, like pulse taking, auscultation, and electrocardiography.

4. Heart rate variability

The interaction of the organisms with its environment causes changes in the metabolic requirements of the multiple tissues that are depending on the circulating blood to supply oxygen and nutrients and remove metabolic waste, i.e., age, physical conditioning or exercise, behavior (emotions, pathologies, spice that is being studied, activity that takes place when the HR is taken, hemorrhages, heart attacks, addictions). In response to these demands, the heart adapts its interbeat intervals.

These intervals vary thanks to the intrinsic properties of the heart (spontaneous activity of the sinoatrial node [6] and atrial and ventricular electrical properties along with extracellular matrix composition) and especially the influence of the autonomic nervous system (ANS), a communication pathway between the heart and the whole body. This system modulates the spontaneous activity of the sinoatrial node and conduction system of the heart (**Figure 2**).

The ANS regulates heart rate, visceral activities, and glandular functions to keep homeostasis. The ANS innervation on the heart can be divided in sympathetic (SNS) and parasympathetic (PNS) nervous system. They both have opposing effects on the heart activity. The sympathetic nervous system is responsible for the "fight or run" response, increasing the myocardium contractile properties and the rate of spontaneous activity of the sinoatrial node (SAN), the natural pacemaker of the heart, augmenting the heart rate. On the other hand, the parasympathetic nervous system has an inhibitory effect on the peacemaker and atrioventricular node (NAV) activity (see **Figure 2**), adjusting to rest states by means of a decrease in the heart rate [2].

Sympathetic innervation secretes norepinephrine, a neurotransmitter that links to β1 receptors on the cardiac sarcolemma activating G proteins. This union induces a conformational change that dissociates the α_s subunit activating adenylyl cyclase. The activated adenylyl cyclase catalyzes the conversion of ATP to AMP_c, which joins directly to ionic channels responsible for the hyperpolarization activated pacemaker

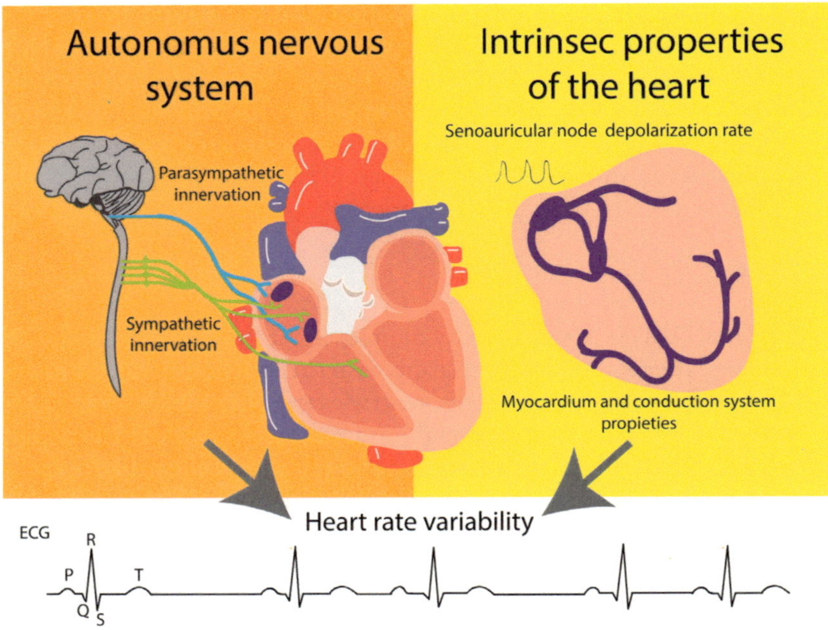

Figure 2.
Heart rate variability sources. The time interval variations between consecutive heartbeats are result of the interaction between the autonomous nervous system modulation and intrinsic properties of the heart regulating its function.

current (I_f) increasing the SAN depolarization rate. This stimulus also increases the opening probability and the inward calcium current enhancing the strength of cardiac contraction. Parasympathetic innervation releases acetylcholine, a neurotransmitter that binds to M_2 receptors on the cardiac sarcolemma activating inhibitory G proteins, inducing a conformational change in G_i protein that dissociates the α_i subunit inhibiting adenylyl cyclase leading to a decrease in the formation of AMP_c, thus decreasing the SAN depolarization rate. The dynamical interaction between SNS and PNS enables the heart to fulfill the organism requirements in the short and long term.

Since the heartbeat is a cyclic phenomenon that repeats continually as a result of the interaction between spontaneous SAN activity [7], passive and active properties of the myocardium, conduction system, and ANS influence, it can be regarded as a result of the interaction of multiple coupled systems that oscillate. This complex nonlinear interaction reflects on interbeat interval variability; such phenomenon is called heart rate variability.

The interbeat intervals are usually assessed as the time between the R-wave peaks of the ECG signal (RR time series). This registry is consequence of the spatial and temporal sum of the electrical activity of the whole heart, and each wave is characteristic of specific electrical events. The R wave is representative of the QRS complex, which is the result of the ventricular transmural depolarization heterogeneity [8]. In view on the fact that the time from the start of the depolarizing wave at the sinoatrial node to the ventricle depolarization can account as other oscillation sources, inter-beat interval indicated by the PP interval (depolarization of the atria) can provide some insights that can be concealed by the RR interval (**Figure 2**).

HRV analysis is a valuable noninvasive method to quantify modifications caused by aging, disease progression, and other physiologic or pathologic changes. These alterations influence the oscillating systems or the way they couple, as sympathetic and parasympathetic heartbeat modulation besides intrinsic properties of the heart that rely on extracellular matrix, sarcolemma composition (ionic channel density and kinetics, gap junction density, lipid composition), myocyte size, adipocyte, and fibroblast distribution. The etiology of these alterations is often related to metabolic diseases [9].

5. Poincaré plots

5.1 Time series

To analyze HRV using the ECG signal, we used the R waves of the QRS complex. The evaluation of the time between an R-wave peak (R_1) and the next immediate R-wave peak (R_2), the time interval between the appearance of an R wave and the next (t_{1-2}) will be called heart period. The RR intervals are organized in chronological order, with an organized set of numbers. This set will be called the "heart activity time series."

The heart rate (number of beats per unit of time) can be estimated as the inverse of the time period. When the frequency is stable, it is always the same, so are the period and the time series. When the time series is plotted against its order appearance, a time series graph is obtained [10]. The times series values determine the shape of the graph. When the frequency is constant, the graph is a parallel line to the time axis. And in the case that the frequency has variations (HRV), the graph is like in **Figure 3**.

The time series has all the information of the variability of system; then, to determine that two time series are similar, numerical values were allocated to this variability. The first tool used was RR time series spectral analysis, this technique is based on the use of all periodic signals consisting of sums of sine and cosine functions with different frequencies and amplitudes, with the purpose to determine which frequencies are involved in the formation of the time series [10]. The frequencies obtained by this mathematical tool have been associated to the nervous system, breathing, and other physiological functions. The frequencies and power spectrum of the different components of the time series are the parameters used to quantify the variability with this method [11].

The disadvantage (if it can be considered as one) of using this method is that not everybody is expert in Fourier series; therefore it's difficult to analyze, interpret, and perform. The second tool we use is the Poincaré plots. They require graphing

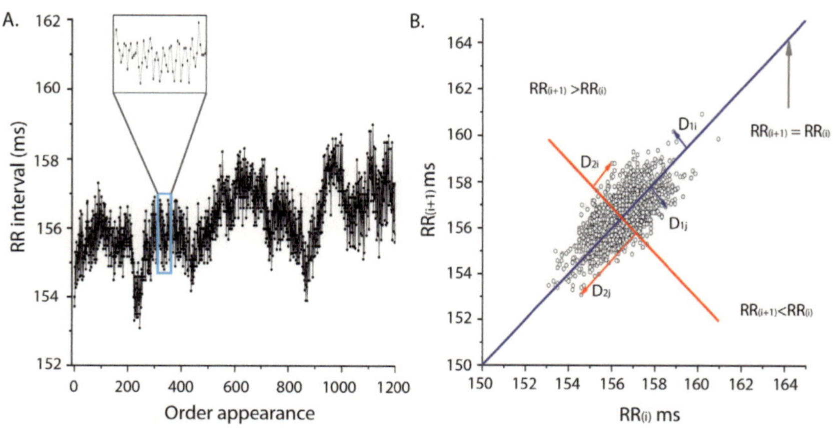

Figure 3.
Heart rate variability analysis. (A) Time series with chronological order and (B) Poincaré plot of time series.

the time series as follows: the first RR_1 interval is assigned as x value, and then the value of the RR_2 interval is assigned as y; this ordered pair is plotted on a Cartesian coordinate axis. Now we consider RR_2 as x, and RR_3 as y to plot it. After that the rest of the RR intervals are graphed in the same way, where the x is the $RR_{(i)}$ interval and the y is the interval $RR_{(i+1)}$; we plot each RR interval against the next immediate one. The resulting graphs are converted into spot stains; this chart is known as Poincaré plot (**Figure 3**). Now the question is how to quantify the spots. Before we give an answer to the problem, we will first describe the advantages of Poincaré plot.

5.2 Advantages of the Poincaré plots

First of all, it is important to mention that the heart rate will be $1/RR_i$; consequently the analysis of the interval variations gives us information of the heart rate variation. That is, the Poincaré plots give us information about changes in heart rate even if this parameter is not explicitly represented in this graph. As above mentioned, the frequency (F) is the inverse of the time period (T), hence $F^*T = 1$.

In the plot we trace the identity straight line $RR_{(i+1)} = RR_{(i)}$, this line divides the plane into three parts: one where the $RR_{(i+1)}$ is equal to $RR_{(i)}$ (blue line in **Figure 3**), another where $RR_{(i+1)} > RR_{(i)}$ which is the top of the identity line. And the third where $RR_{(i+1)} < RR_{(i)}$ which is the part that is under the identity line (**Figure 3**). Therefore, just by looking at the point localization, we can say that the next interval has a higher value, i.e., the frequency is less. In other words, when the points fall above the identity straight line, the period i + 1 is greater and the further the point is from the identity line, the value of the period i + 1 will be greater; otherwise when the points are under the identity graph, the period i + 1 will be smaller, and the frequency will be higher [3].

5.3 SDD_1 and SDD_2 calculation

The distance between the points and the identity straight line tells us what the instantaneous (or sequential) changes of the RR interval will look like, as mentioned in the short term [12]. As an example, we will mention that when the distance from the points to the identity straight line is zero $RR_{(i+1)} = RR_{(i)}$, there are no changes in the interval, but if this distance becomes greater, the variation between RR is greater. These distances are called D_{1i}; the D_{1i} distances that are above the identity straight line will be positive and those below will be negative in such a way that the average of these distances are zero, but the standard deviation of these distances (SDD_1) will be different from zero, and this parameter will be used to characterize the width of the Poincaré plot. The width or SDD_1 will be used to determine the variability of D1; this parameter is related to the short-term variability of the RR, and this relates to the interaction of the sympathetic system and the heart. To calculate SDD_1 all distances from points to the identity line are calculated the average and standard deviation [3, 13]. Thus it is found that

$$D_{1i} = \sqrt{\left(\frac{RR_i - RR_{i+1}}{2}\right)^2} \tag{1}$$

All D_{1i} distances are added and divided by the number of distances to get the average; the standard deviation to the latter is called SDD_1. SDD_1 is a parameter that characterizes short-term variability.

Secondly, calculate the distance from all points to the perpendicular line that crosses the identity line at the coordinate point of the mean value (RR_m, RR_m). This distance is called $D2_j$ (Eq. (2) and **Figure 3**). All D_{2j} distances are added and divided by the number of distances to get the average; the standard deviation to the latter is called SDD_2.

The RR variability changes can be obtained based on the RR time series without using explicitly the time:

$$D2j = \sqrt{2\left(\frac{2\overline{RR_j} - RR_j - RR_{j+1}}{2}\right)^2} \quad (2)$$

where $\overline{RR_j}$ is the average value of the sum of all RR intervals.

By obtaining the values of SDD_1 and SDD_2, we quantify the variability of the heart rate in the short and long term. This data defines the coefficient of variability as SDD_1/SDD_2.

Using the Poincaré plots, the quantification of the variability in the heart rate is determined by calculating SDD_1, SDD_2, and the SDD_1/SDD_2 ratio. The advantage of this method is that the calculation of these parameters is clearly arithmetic, and just by looking at the Poincaré, plot you have an idea of how the variability is given.

6. Biomarkers

A biological marker or biomarker is any substance, structure, or process that is objectively measured and evaluated as an indicator of normal biological processes. The biomarkers in the medical science field play essential role for disease detection, pathogenic responses, and therapeutic intervention. These markers are observational side products with potential utility in clinical and research studies [14]. Additionally they are used in new treatment strategies for clinical management. The biomarker field opens the opportunity to originate new knowledge in the complex health scheme.

7. Metabolic disease

Metabolic alterations cause metabolic diseases as result of changes in chemical reactions in the organism by several enzyme deficit, developing alterations like lipid metabolism disorders. These diseases are associated with synthesis and degradation of fatty acids. The principal and general symptoms of metabolism injury are lethargy, weight alteration, inflammatory process, seizures, and jaundice.

8. Metabolic syndrome

For the last decade, the cardiovascular diseases have been the first cause of death worldwide, and the deadly arrhythmias have increased in the industrialized countries; this fact is related to lifestyle and metabolic alterations such as sedentarism and diet [15, 16]. Obesity and metabolic syndrome are disorders associated with metabolic modifications.

The metabolic syndrome has been described as a cluster of several signs like abdominal obesity, hyperglycemia, dyslipidemia, and high blood pressure (**Figure 4**). These factors predispose to develop cardiovascular diseases, and each component is strongly correlated with CVD. An opportune diagnosis is necessary to know the progression of MeS and predisposition to develop lethal risks. HRV analysis is a tool to assess cardiac function in patients with several pathologic conditions.

However, relationships between HRV and cardiac rhythm with changes in MeS have not been found, improving considerably the prognostic and diagnostic of MeS, as well as the side effects.

The ECG is a biomarker for early diagnosis of metabolic diseases [3], and to asses HRV, a five random minute interval must be measured and analyzed. When more time is analyzed, the characteristic SDD_1 and SDD_2 will be lost [3]. In humans, the MeS showed changes in RR intervals; SDD_1 or short-term variability was modified in young adults, while in woman and elderly human, the alterations were vagal as sympathovagal balance (SDD_1 and SDD_2 [17]).

Also, spectral analysis with Fourier transform was used for the 24-h ECG record; this analysis showed that in human, the high frequencies (HF 0.15–0.40 Hz), which represents sympathetic modulation, were lower only in women with metabolic syndrome [18] and at low frequencies (LF 0.04–0.15 Hz), which represent parasympathetic modulation, heart rate was not altered by MeS. Furthermore, individual components of the MeS were highly correlated with imbalance cardiac autonomic system; the obesity modifies sympathetic nervous system [19]; hyperglycemia alters parasympathetic system [20]; and microalbuminuria, dyslipidemia, and hypertension do not alter neither of them but decrease LF/HF index (see **Figure 4**) [21, 22].

Rats with obesity and hypertension presented similar cardiovascular changes as humans: a decrease in parasympathetic system without any increase in sympathetic modulation [23], and only temporary alterations in sympathetic nervous system were reported in rats with high sucrose diet, insulin resistance, and visceral fat (epididymal fat) [24]. However, the rats with high sucrose diet showed higher LF than control [25], and also the heart rate was decreased showing sinus bradycardia and a threefold increase of heart rate variability, SDD_1 15 ± 0.4, and SDD_2 69 ± 1, compared with control animals 5.5 ± 0.1 and 26 ± 0.1, respectively. In addition, sinoatrial node doubled its variability as shown in the SDD_1/SDD_2 index = 0.25 for control condition and MeS: SDD_1/SDD_2 = 0.55 [26]. In genetically modified rats, cardiac alterations were observed independently on individual characteristic of MeS (see **Figure 4**).

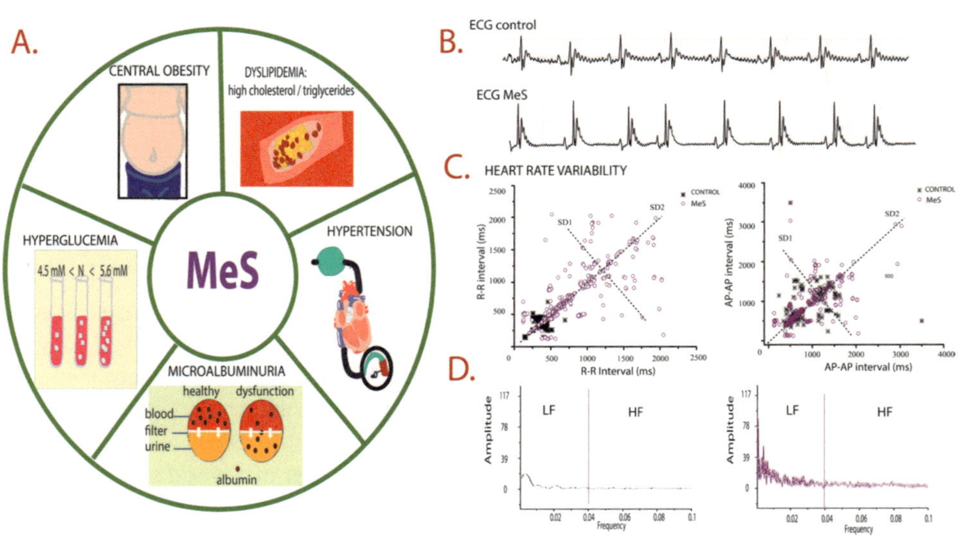

Figure 4.
The metabolic syndrome increases the heart rate variability. (A) The cluster signs of MeS increase the risk to develop cardiovascular diseases (CVD) and diabetes mellitus. (B) ECG of control and MeS rats, showing lower heart rate in MeS rats. (C) Poincaré plots exhibiting lower balance between parasympathetic and sympathetic systems. (D) Fourier analysis indicating that lower frequencies predominate in MeS rat RR time series.

9. Diabetes mellitus

The cardiac arrhythmias in diabetes mellitus are due to structural and functional remodeling, which are alterations in the architecture of the heart that include fibrosis, fat deposition, hypertrophy, modification in the utilization, and production of energy. In addition, electrical activity remodeling includes failure in electrical conduction, dysregulation in ion channels and gap junctions [27], and all these changes are added to the autonomic imbalance between the sympathetic and parasympathetic nervous systems until it becomes cardiac autonomic neuropathy (CAN), which is recognized as a risk for development of atrial fibrillation and sudden cardiac death (see **Figure 5**) [28].

In order to realize clinical diagnosis of CAN, the performing cardiac autonomic reflex test or neuropathy Ewing's battery is recommended. It consists of the assessment of the HRV in rest condition, while standing, during paced deep breathing, during sustained muscle contraction with the use of a handgrip dynamometer (handgrip exercise), and during and after a provoked increase in intrathoracic/abdominal pressure (maneuver of Valsalva) (see **Figure 5**) [29, 30].

Unfortunately, these tests have limitations: patients must be aware so they can perform each of the tests, and it is necessary to suspend medications that could alter the results of the test (e.g., the avoidance of medications that cause hypotension, such as diuretics, tricyclic antidepressants, and vasodilators) [31].

Due to these disadvantages, the measurement of HRV has been used as an alternative for CAN diagnosis in recent years because it is a noninvasive test, it does not provoke pain in the patient, the analysis is performed in a short time, it is reliable, and it is a low-cost technique. In addition, this methodology allows the HRV analysis to be performed in less time because it is not necessary to have specialized knowledge in statistics or mathematics since the values of SDD_1, SDD_2, and SDD_1/SDD_2 are obtained by means of relatively simple arithmetic calculations and it does not need specialized software to perform them [3]. Another improvement is that the Poincaré plot analysis can be done with only 100 RR intervals, which excludes the use of a Holter registry without reducing the reliability and sensitivity of the test [32].

Several authors have reported a decrease in HRV in patients with DM types 1 and 2 regardless of the method used to measure it (frequency-domain HRV or time domain). The decrease in HRV in diabetic patients is associated with an early phase of the evolution of CAN. There is a loss of parasympathetic function with a relative increase of sympathetic function causing an imbalance of the sympathetic/parasympathetic tone (without parasympathetic denervation). The patient experiments an increase in resting heart rate. In the next stage, sympathetic denervation takes place increasing the risk of arrhythmias [33]. Despite the existence of a large number of studies on HRV in diabetic patients, we still do not have a relationship that allows us to know the stage of damage in which the autonomic nervous system is found.

On the other hand, we have validated the use of HRV and the measurement of SDD_1, SDD_2, and the Poincaré SDD_1/SDD_2 index (Eqs. (1) and (2)) as a biomarker for diagnosis and prognosis of cardiac autonomic neuropathy. For this purpose, a model of type 1 diabetes pharmacologically induced by STZ was used. This model was developed in CD1 mice in which the progress of disease is allowed for 10 weeks without insulin administration (the time compared with human 8 years of disease progression), which produces a decrease in the values of SDD_1 (1 vs. 0.9), SDD_2 (1.3 vs. 0.8), and SDD_1/SDD_2 (0.8 vs. 1.1) compared to the control (**Figure 6**) [32].

In this stage of the disease, no decrease in heart rate was reported, which suggests that CAN was in the early stages. However, after a time period equivalent to

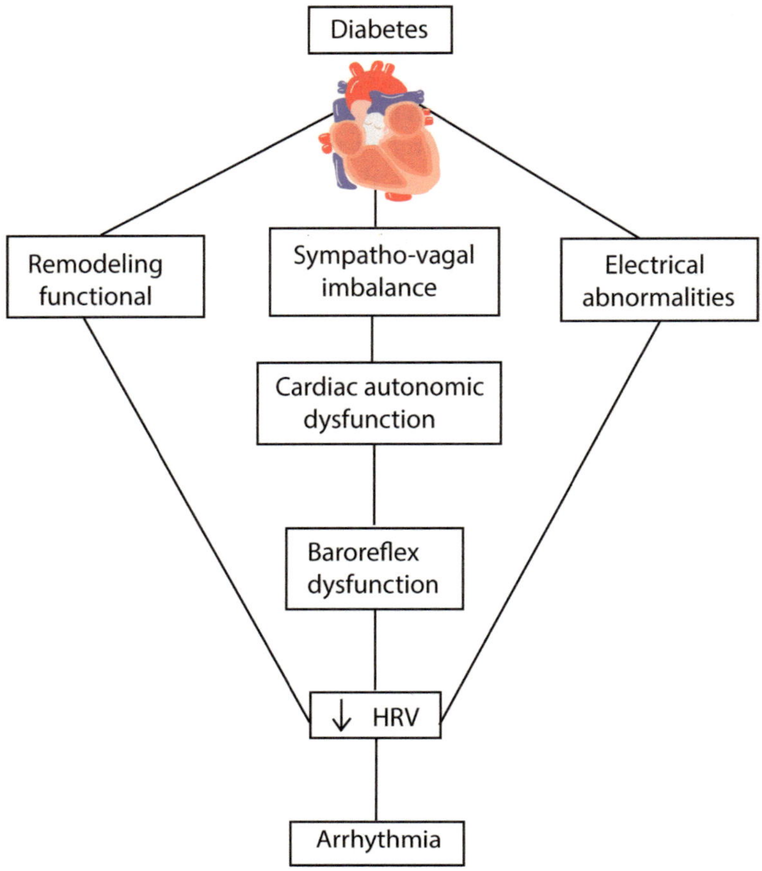

Figure 5.
Relations between alterations in DM and cardiac arrhythmias. The alterations in the architecture of heart tissue and functions produce a decrease in HRV during diabetes, which increment the risk of arrhythmias.

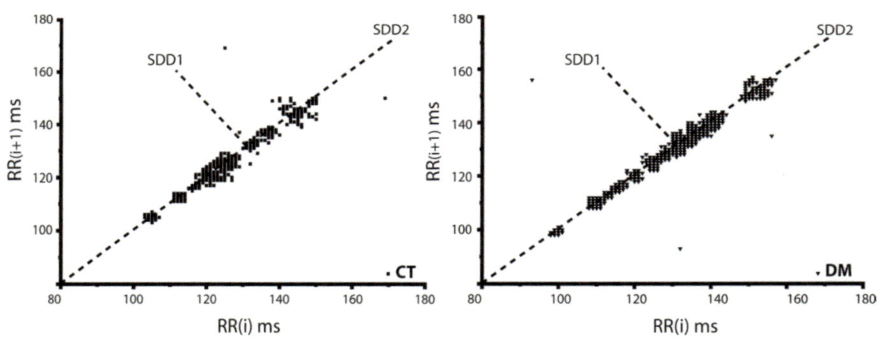

Figure 6.
Poincaré plots of interval RR of ECG. The HRV in conditions of control vs. 8 years of DM development. The influence of the ANS allowed maintaining the balance of an elliptical shape.

15 human years of DM induction without hypoglycemic treatment CAN, a decrease in HRV is developed (**Figure 6**). Additionally, in this second stage, mice showed denervation in the pacemaker tissue [13]. We conclude that the use of HRV and Poincaré plots could detect CAN even in early stages of the disease, and therefore it will allow introducing therapeutic maneuvers to control the symptoms and delay the damage to the ANS due to DM.

10. Conclusions

The periodic oscillations in biological phenomena are quantified with the purpose to use them as a health indicator (biomarker) in mammalian. By means of the ECG interval analysis, HRV is quantified using RR and PP time series. Poincaré plots were constructed, and three indicators were obtained: SDD_1, SDD_2, and SDD_1/SDD_2 index. The behavior of these indicators is related with health or metabolic disease. In MeS, a sympathovagal imbalance was reported, and the parasympathetic system showed alterations with a twofold increase in SDD_2 indicator. Furthermore, the three indicators were decreased by DM. These biomarkers have the advantages of being based on a noninvasive tool, being objective, and being obtained by easy arithmetic calculus. In addition, the shape of the Poincaré plots offers qualitative information by only looking at it.

Author details

Alondra Albarado-Ibañez[1], Rosa Elena Arroyo-Carmona[2], Daniela Alexandra Bernabé-Sánchez[1], Marissa Limón-Cantú[1], Benjamín López-Silva[1], Martha Lucía Ita-Amador[3] and Julián Torres-Jácome[1*]

1 Physiology Institute, Benemérita Universidad Autónoma de Puebla, Puebla, Mexico

2 Faculty of Chemistry Sciences, Benemérita Universidad Autónoma de Puebla, Puebla, Mexico

3 Nororiental Complex, Faculty of Medicine, Benemérita Universidad Autónoma de Puebla, Puebla, Mexico

*Address all correspondence to: jtorresjacome@gmail.com

IntechOpen

© 2019 The Author(s). Licensee IntechOpen. This chapter is distributed under the terms of the Creative Commons Attribution License (http://creativecommons.org/licenses/by/3.0), which permits unrestricted use, distribution, and reproduction in any medium, provided the original work is properly cited.

References

[1] Dardente H, Wood S, Ebling F, de Miera CS. An integrative view of mammalian seasonal neuroendocrinology. Journal of Neuroendocrinology. 2019;**31**(5):1-17

[2] Gordan R, Gwathmey JK, Xie LH. Autonomic and endocrine control of cardiovascular function. World Journal of Cardiology. 2015;**7**(4):204-214

[3] Arroyo-Carmona RE, Lopez-Serrano AL, Albarado-Ibanez A, Mendoza-Lucero FM, Medel-Cajica D, Lopez-Mayorga RM, et al. Heart rate variability as early biomarker for the evaluation of diabetes mellitus progress. Journal Diabetes Research. 2016;**2016**:8483537

[4] Aschoff J. Activity in anticipation and in succession of a daily meal. Bollettino della Società Italiana di Biologia Sperimentale. 1991;**67**(3):213-228

[5] Sharma VK, Chandrashekaran MK. Zeitgebers (time cues) for biological clocks. Current Science. 2005;**89**(7):1136-1146

[6] Yaniv Y, Ahmet I, Liu J, Lyashkov AE, Guiriba TR, Okamoto Y, et al. Synchronization of sinoatrial node pacemaker cell clocks and its autonomic modulation impart complexity to heart beating intervals. Heart Rhythm. 2014;**11**(7):1210-1219

[7] Bergfeldt L, Haga Y. Power spectral and Poincare plot characteristics in sinus node dysfunction. Journal of Applied Physiology (Bethesda, MD: 1985). 2003;**94**(6):2217-2224

[8] Antzelevitch C, Shimizu W, Yan GX, Sicouri S. Cellular basis for QT dispersion. Journal of Electrocardiology. 1997;**30**:168-175

[9] Acharya UR, Faust O, Sree SV, Ghista DN, Dua S, Joseph P, et al. An integrated diabetic index using heart rate variability signal features for diagnosis of diabetes. Computer Methods in Biomechanics and Biomedical Engineering. 2013;**16**(2):222-234

[10] Singh D, Vinod K, Saxena SC. Sampling frequency of the RR interval time series for spectral analysis of heart rate variability. Journal of Medical Engineering & Technology. 2004;**28**(6):263-272

[11] Malliani A, Lombardi F, Pagani M. Power spectrum analysis of heart-rate-variability-a tool to explore neural regulatory mechanisms. British Heart Journal. 1994;**71**(1):1-2

[12] Brennan M, Palaniswami M, Kamen P. Do existing measures of Poincare plot geometry reflect nonlinear features of heart rate variability? IEEE Transactions on Biomedical Engineering. 2001;**48**(11):1342-1347

[13] Albarado-Ibanez A, Arroyo-Carmona RE, Sanchez-Hernandez R, Ramos-Ortiz G, Frank A, Garcia-Gudino D, et al. The role of the autonomic nervous system on cardiac rhythm during the evolution of diabetes mellitus using heart rate variability as a biomarker. Journal of Diabetes Research. 2019;**2019**:10

[14] Strimbu K, Tavel JA. What are biomarkers? Current Opinion in HIV and AIDS. 2010;**5**(6):463-466

[15] Tadic M, Ivanovic B, Celic V, Cuspidi C. Are the metabolic syndrome, blood pressure pattern, and their interaction responsible for the right ventricular remodeling? Blood Pressure Monitoring. 2013;**18**(4):195-202

[16] Isik T, Kurt M, Uyarel H. Metabolic syndrome without overt diabetes is associated with prolonged pro-arrhythmogenic electrocardiographic parameters. Anadolu Kardiyoloji Dergisi. 2012;**12**(6):529

[17] Xhyheri B, Manfrini O, Mazzolini M, Pizzi C, Bugiardini R. Heart rate variability today. Progress in Cardiovascular Diseases. 2012;**55**(3):321-331

[18] Camm AJ, Malik M, Bigger JT, Breithardt G, Cerutti S, Cohen RJ, et al. Heart rate variability. Standards of measurement, physiological interpretation, and clinical use. European Heart Journal. 1996;**17**(3):354-381

[19] Min KB, Min JY, Paek D, Cho SI. The impact of the components of metabolic syndrome on heart rate variability: Using the NCEP-ATP III and IDF definitions. Pacing and Clinical Electrophysiology. 2008;**31**(5):584-591

[20] Stein PK, Barzilay JI, Domitrovich PP, Chaves PM, Gottdiener JS, Heckbert SR, et al. The relationship of heart rate and heart rate variability to non-diabetic fasting glucose levels and the metabolic syndrome: The Cardiovascular Health Study. Diabetic Medicine. 2007;**24**(8):855-863

[21] Liao DP, Carnethon M, Evans GW, Cascio WE, Heiss G. Lower heart rate variability is associated with the development of coronary heart disease in individuals with diabetes: TThe Atherosclerosis Risk in Communities (ARIC) Study. Diabetes. 2002;**51**(12):3524-3531

[22] Liao DP, Sloan RP, Cascio WE, Folsom AR, Liese AD, Evans GW, et al. Multiple metabolic syndrome is associated with lower heart rate variability: The atherosclerosis risk in communities study. Diabetes Care. 1998;**21**(12):2116-2122

[23] Grassi G, Arenare F, Quarti-Trevano F, Seravalle G, Mancia G. Heart rate, sympathetic cardiovascular influences, and the metabolic syndrome. Progress in Cardiovascular Diseases. 2009;**52**(1):31-37

[24] da Silva RJ, Bernardes N, Brito JD, Sanches IC, Irigoyen MC, De Angelis K. Simvastatin-induced cardiac autonomic control improvement in fructose-fed female rats. Clinics. 2011;**66**(10):1793-1796

[25] Albarado-Ibañez A, Hiriart M, Hoeflich AF, Torres-Jácome J. Assessment the change on rhythm cardiac produces by the metabolic syndrome in rats: Using nonlinear methods. In: Conference Nonlinear. 2018;**1**(1):2016-6

[26] Albarado-Ibanez A, Avelino-Cruz JE, Velasco M, Torres-Jacome J, Hiriart M. Metabolic syndrome remodels electrical activity of the sinoatrial node and produces arrhythmias in rats. PLoS One. 2013;**8**(11):e76534

[27] Grisanti LA. Diabetes and arrhythmias: Pathophysiology, mechanisms and therapeutic outcomes. Frontiers in Physiology. 2018;**9**:1669

[28] Agarwal SK, Norby FL, Whitsel EA, Soliman EZ, Chen LY, Loehr LR, et al. Cardiac autonomic dysfunction and incidence of atrial fibrillation results from 20 years follow-up. Journal of the American College of Cardiology. 2017;**69**(3):291-299

[29] Fisher VL, Tahrani AA. Cardiac autonomic neuropathy in patients with diabetes mellitus: Current perspectives. Diabetes, Metabolic Syndrome and Obesity: Targets and Therapy. 2017;**10**:419-434

[30] Bissinger A. Cardiac autonomic neuropathy: Why should cardiologists care about that? Journal of Diabetes Research. 2017;**5374176**:1-9

[31] Kuehl M, Stevens MJ. Cardiovascular autonomic neuropathies as complications of diabetes mellitus. Nature Reviews Endocrinology. 2012;**8**(7):405-416

[32] Nussinovitch U, Cohen O, Kaminer K, Ilani J, Nussinovitch N. Evaluating reliability of ultra-short ECG indices of heart rate variability in diabetes mellitus patients. Journal of Diabetes and its Complications. 2012;**26**(5):450-453

[33] Vinik E, Silva MP, Vinik AI. Measuring the relationship of quality of life and health status, including tumor burden, symptoms, and biochemical measures in patients with neuroendocrine Tumors. Endocrinology and Metabolism Clinics. 2011;**40**(1):97

Chapter 7

Evolution of Parasympathetic Modulation throughout the Life Cycle

Moacir Fernandes de Godoy and Michele Lima Gregório

Abstract

Based on the largest data set ever available for analysis of heart rate variability (HRV) variables, in healthy individuals, it was possible to determine the evolutionary behavior of three representative components of parasympathetic nervous system function (RMSSD, PNN50, and HF ms^2), in different age groups of the life cycle: newborns, children and adolescents, young adults, and, finally, middle-aged adults. A near-parabolic and nonsynchronous behavior was observed among the different variables evaluated, with low values at first, then progressive elevation, and later fall, approximating the values of the newborns to the values of middle-aged adults and suggesting that the autonomic nervous system, at least relatively to its parasympathetic component, undergoes a growing maturation that is completed in the young adult and later suffers a progressive degeneration, completing the life cycle. This fact should be considered when comparing the analysis between healthy individuals and those with different states of pathological impairment.

Keywords: autonomic nervous system, parasympathetic nervous system, heart rate variability, homeostasis, life cycle

1. Introduction

The autonomic nervous system (ANS) is a division of the peripheral nervous system and, based on anatomy and physiology, has three subdivisions: sympathetic nervous system (SNS), parasympathetic nervous system (PNS), and enteric nervous system (ENS). SNS has thoracolumbar distribution, and PNS has a craniosacral distribution, while ENS is the major part of the peripheral nervous system being found throughout the gastrointestinal tract, extending from the esophagus to the rectum, and is also present in the pancreas and in the gallbladder [1–4].

ANS has the responsibility to ensure that homeostasis be maintained in the face of disturbances produced by both the external and internal environment [5]. In the heart of rats, ANS begins its development on the embryonic 18.5 day until the twenty-first postnatal day (P21) [6].

Sympathetic neurons are located in the paravertebral ganglia, have long axonal projections to the organs, and produce excitatory effects mediated by the noradrenergic transmitter norepinephrine (NE). Conversely, parasympathetic neurons are located in ganglia near or on the surface of organs, have shorter axonal projections, and produce inhibitory effects mediated by the cholinergic transmitter

IntechOpen

acetylcholine (ACh). The enteric nervous system provides the intrinsic innervation of the gut, controlling different aspects of the gut function, such as motility [4].

Although ANS can actually function autonomously, the central nervous system can contribute to a significant regulatory effect [3].

Heart rate variability (HRV) analysis is a practical, noninvasive, reproducible, and cost-effective resource that has been widely applied to study the autonomic behavior of the human organism, being particularly useful for the evaluation of sympathetic and parasympathetic components, although with regard to sympathetic behavior, there is still controversy about the mechanisms involved [7].

Higher vagally mediated heart rate variability is associated with better autonomic balance, better health outcomes, and flexible physiological responses. In contrast, lower HRV is associated with disease and all-cause mortality [8].

In [9], some reference values for normality of HRV variables are suggested, although highlighting that "As no comprehensive investigations of all HRV indices in large normal populations have yet been performed, some of the normal values [...] were obtained from studies involving small number of subjects."

The reference values for normality cited and recommended in the Task Force were taken from the work of Bigger et al. (1995). The authors were based on only 274 individuals considered healthy and restricted to be 40–69 years old [10].

The aim of this chapter is restricted to the parasympathetic division of ANS. For the evaluation of this component, there is a well-established consensus that some variables, such as the root mean square of the successive RR interval differences (RMSSD), the percent of normal RR intervals that differed by more than 50 ms (PNN50) both in the time domain, and the absolute power of the high-frequency band component (HF ms^2), in the frequency domain, specifically represent vagal modulation, presenting both diagnostic and prognostic properties [11–12].

Generally speaking, heart rate variability analysis has become the most used noninvasive tool to evaluate autonomic control mechanisms and to predict mortality risk in several clinical conditions, including coronary artery disease, heart failure, diabetes, and hypertension [13].

According to Goldberger et al. [14], there was some evidence that age influenced the responsiveness of the HRV parameters with changing parasympathetic effect. They studied 29 normal volunteers (15 women; mean age 39 $\pm$ 12 years) after β-adrenergic blockade with intravenous propranolol. Five-minute ECG recordings were made during graded infusions of phenylephrine and nitroprusside to achieve baroreflex-mediated increases and decreases in parasympathetic effect, respectively. There was some evidence that age influenced the responsiveness of the HRV parameters with changing parasympathetic effect, with significant association for RMSSD and PNN50.

Despite the significant amount of studies in the literature dealing with the HRV and autonomic regulation subject, there is a lack of studies with large series, addressing several variables in different age ranges, from birth to the elderly adult. So, we will evaluate the contribution of these three variables in the study of parasympathetic autonomic behavior throughout the life cycle based on the evaluation of a significant amount of data (835,902 in total) extracted from the literature regarding heart rate variability variables and admittedly related to the parasympathetic nervous system being 53,882 results from healthy individuals.

2. Method

The inclusion criterion was quite broad in view of the proposed objective, which was to establish reference values, based on the largest amount of information possible. Thus, by searching the available databases (PubMed, Google Scholar, Cochrane Library, ScienceDirect, Wiley Online Library, SciELO, LILACS, and Thesis Banks of Brazilian Universities, among others) and following the PRISMA 2009 flow diagram [15], articles evaluating the values of heart rate variability (**Flow Diagram**) were included, and after, those directly related to the parasympathetic component of ANS, in the time domains (RMSSD and PNN50) and in the frequency domain (HF ms^2), in humans, regardless of age and gender and also regardless of the length of the time series, patient position, and analysis equipment, were selected but provided that the data were always collected from individuals specifically considered to be healthy. Based on this criterion, it is noteworthy that the individuals, who in the original work were cataloged as being from the general population, were not considered to be healthy because there are known comorbidities in this type of sample, and so, they were not included.

Values with evident evidence of extreme outliers (three or more standard deviations below the first quartile or above the third quartile, from the set of values collected for a given variable) were excluded.

Flow diagram

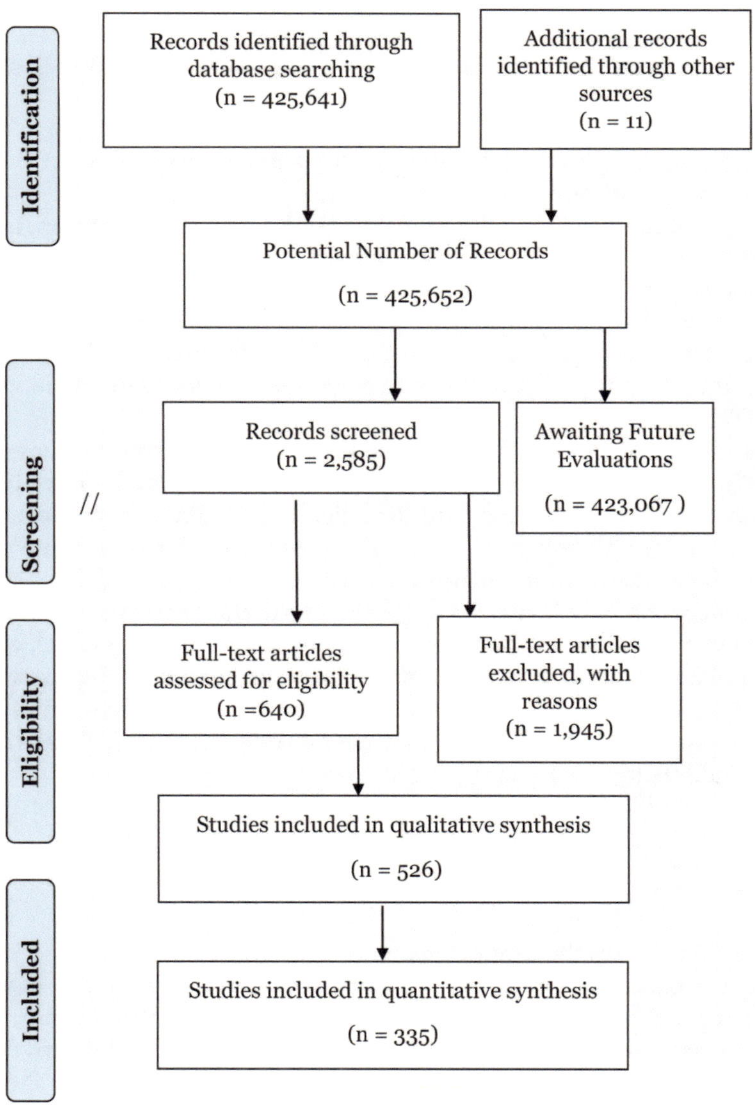

Domain	Variable	Total group	General population + diseased	Healthy
Time	RMSSD ms	208,657	183,155	25,502
Time	PNN50	49,400	35,043	14,357
Frequency	HF ms2	159,894	145,871	14,023

Table 1.
Distribution of the literature data evaluated, in terms of the variable studied, highlighting the sample of interest (healthy individuals) and its size in relation to the total amount obtained.

Age range (years)	Age mean ± SD	RMSSD (ms)	PNN50 (%)	HF(ms^2)
Newborns	[0 a 3 days]	234	78	272
Up to 20	13.29 ± 4.64	4,419	2,790	4,346
20–40	25.21 ± 4.88	8,459	1,031	5,721
40–70	52.74 ± 7.56	12,390	10,468	3,684
Totals		25,502	14,357	14,023

Table 2.
Mean and standard deviation of the analyzed age groups and respective amounts of data analyzed, by studied variable..

Table 1 informs the studied variable, its domain, and the amount of values collected in the literature.

RMSSD (root mean square of the successive RR intervals differences, in ms; PNN50 (percent of normal RR intervals that differed by more than 50 ms in %); HF (absolute power of the high-frequency band; 0.15–0.40 Hz, in ms^2).

Groupings were made by age range to precisely characterize the evolutionary behavior of the parasympathetic system throughout the life cycle. The amounts of data evaluated for each group and their average ages and standard deviations are shown in **Table 2**.

From all included studies, the mean and the standard deviation values of each variable of interest were extracted. The overall mean value was obtained by weighted average. The global standard deviation was obtained from the individual mean set of each study. As the collected values were the means and standard deviations, the existence of normality was assumed. The values from the different age groups were compared with the aid of the unpaired t-test assuming that the standard deviations of each group were not similar to each other (Welch correction). GraphPad InStat version 3.00 software was used to obtain P-values. A PDF file containing all the 335 references used to mounting the database can be solicited to the correspondent author. The large number of references would make it impossible to include them directly in the present text.

3. Results

Table 3 summarizes the results obtained.

RMSSD (root mean square of the successive RR intervals differences in ms; PNN50 (percent of normal RR intervals that differed by more than 50 ms), HF (absolute power of the high-frequency band; 0.15–0.40 Hz); SD, standard deviation.

Group	Age range	RMSSD	PNN50	HF
		Mean ± SD	Mean ± SD	Mean ± SD
1	Newborns	11.6 ± 0.9	1.4 ± 3.7	66.7 ± 85.5
2	Up to 20	52.0 ± 18.0	25.7 ± 11.6	1124.0 ± 710.8
3	20–40	53.1 ± 22.2	19.9 ± 12.9	2067.2 ± 1144.7
4	40–70	28.2 ± 11.8	6.9 ± 0.3	236.3 ± 248.5

Table 3.
Mean and standard deviation of the variables studied according to the different age groups.

The statistical analysis (p-values, t-test unpaired, two-tailed, Welch correction) comparing the mean values for each variable along the age ranges is showed below.

Group	RMSSD	PNN50	HF
1 versus 2	$P < 0.0001$	$P < 0.0001$	$P < 0.0001$
1 versus 3	$P < 0.0001$	$P < 0.0001$	$P < 0.0001$
1 versus 4	$P < 0.0001$	$P < 0.0001$	$P < 0.0001$
2 versus 3	$P = 0.0024$	$P < 0.0001$	$P < 0.0001$
2 versus 4	$P < 0.0001$	$P < 0.0001$	$P < 0.0001$
3 versus 4	$P < 0.0001$	$P < 0.0001$	$P < 0.0001$

As can be observed, the P-values were extremely robust indicating significant extreme differences for all comparisons.

Figures were constructed showing the behavior of each variable along the progressive increase in chronological age, from the healthy newborn group (subgroup 1) to children and adolescents (subgroup 2) and young adults (subgroup 3), until reaching the middle-aged adults (subgroup 4).

Figures 1–3 graphically demonstrate this behavior.

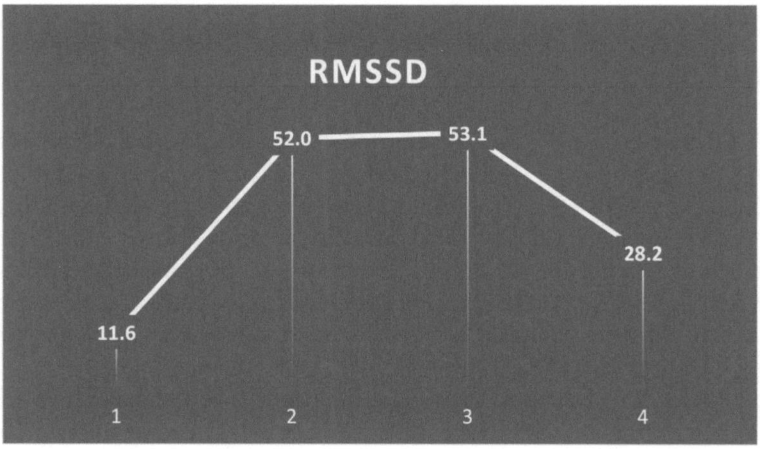

Figure 1.
Mean evolutionary behavior of RMSSS values for the different age groups studied. RMSSD (root mean square of the successive RR interval differences in ms; 1, healthy newborns subgroup; 2, children and adolescents (up to 20 years) subgroup; 3, young adults (20–40 years) subgroup; 4, middle-aged adults (40–70 years) subgroup.

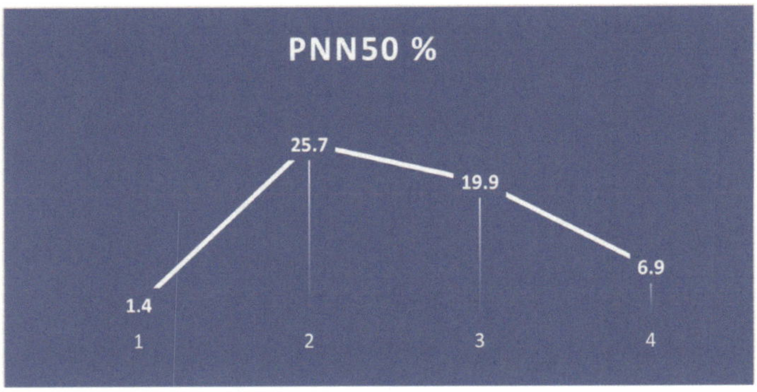

Figure 2.
Mean evolutionary behavior of PNN50 values for the different age groups studied.PNN50% ((percent of normal R-R intervals that differed by more than 50 ms); 1, healthy newborns subgroup; 2, children and adolescents (up to 20 years) subgroup; 3, young adults (20–40 years) subgroup; 4, middle-aged adults (40–70 years) subgroup.

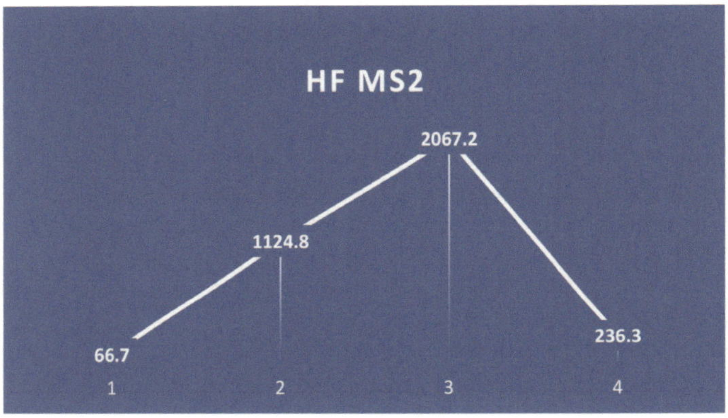

Figure 3.
Mean evolutionary behavior of HF ms² values for the different age groups studied. HF ms² (absolute power of the high-frequency band; 0.15–0.40 Hz); 1, healthy newborns subgroup; 2, children and adolescents (up to 20 years) subgroup; 3: Young adults (20–40 years) subgroup; 4, middle-aged adults (40–70 years) subgroup.

4. Discussion

It is well known that the heart rate variability declines with age. Bonnemeier et al. (2003) [16] obtained 24h recordings from 166 healthy volunteers (85 men and 81 women) aged 20–70 years. They found the most dramatic HRV parameter decrease between the second and third decades. Almeida-Santos et al. (2016) [17] obtained 24h ECG recordings of 1743 subjects of 40–100 years of age. They found a linear decline in SDNN, SDANN, and SDNN index. Curiously, they described U-shaped pattern for RMSSD and pNN50 with aging, decreasing from 40 to 60 and then increasing after age 70.

The present study adds new information about this evolutionary behavior. It was quite clear that parasympathetic autonomic development in healthy individuals is peculiar, being reduced at birth, presenting a progressive elevation up to about 20 years of age (for the three variables studied), and typically, after that initial elevation, two different patterns of behavior occur. The RMSSD variable arises a little more until around 40 years of age when it then begins to decline progressively (**Figure 1**), which we might call as a "'negatively skewed tent' behavior." The PNN50 variable, once reaching its maximum levels around the age of 20, begins to decline progressively until the age of 70 (**Figure 2**), which would graphically be a "positively skewed tent" behavior. Finally, the HF variable rises from birth to about 40 years, when it begins to decline until 70 years of age being graphically a "negatively skewed tent" behavior (**Figure 3**).

We did not find significant studies on heart rate variability in healthy individuals over 70s, probably because above that age, the vast majority of the individuals already have some pathological impairment. Yes, it would exist for the general population, but that was not the focus at this moment. Therefore, a complete definition of HRV behavior in that older group, based on a significant sample like that used here for the other age groups, was not yet possible.

The significant amount of data obtained, together with the extremely significant difference between the values in the different age groups, strongly indicates that this was not a casual finding but a true expression of parasympathetic autonomic behavior.

This is a relevant finding as it sheds new light on the knowledge of normal values in different age groups, since the current gold standard is still established by the Task Force data, based on only 274 cases and exclusively on the age range of 40–69 years.

5. Conclusion

Like every other complex system, in accordance with Chaos Theory, ANS, at least in its parasympathetic component, exhibits a near-parabolic and nonsynchronous behavior for the main variables that evaluates it using heart rate variability, and this fact should be considered in the comparative analysis between healthy individuals and those with different grades of pathological impairment.

Based on the largest data set ever available for healthy individuals, the found values can be proposed as reference standards for future studies about heart rate variability.

Acknowledgements

The authors would like to thank the Brazilian CNPq (National Council for Scientific and Technological Development) [Processes 308759/2015-0 and 308555/2018-0] and to FAPESP (São Paulo Research Foundation) [Process 2017/125297] for the financial support.

Conflict of interest

The authors declare no conflict of interest.

Author details

Moacir Fernandes de Godoy[1,2]* and Michele Lima Gregório[2]

1 Department of Cardiology and Cardiovascular Surgery, Sao Jose do Rio Preto Medical School – Famerp, Sao Jose do Rio Preto, SP, Brazil

2 Transdisciplinary Nucleus for Chaos and Complexity Studies – NUTECC – Sao Jose do Rio Preto Medical School – Famerp, Sao Jose do Rio Preto, SP, Brazil

*Address all correspondence to: mf60204@gmail.com

IntechOpen

© 2019 The Author(s). Licensee IntechOpen. This chapter is distributed under the terms of the Creative Commons Attribution License (http://creativecommons.org/licenses/by/3.0), which permits unrestricted use, distribution, and reproduction in any medium, provided the original work is properly cited.

References

[1] Rang HP, Dale MM, Ritter JM, Flower RJ. Farmacologia. 6th ed. Rio de Janeiro: Editora Elsevier; 2007

[2] Kim C-H, Development KK-S. Differentiation of autonomic neurons. In: Robertson D, Biaggioni I, Burnstock G, Low PA, Paton JFR, editors. Primer on the Autonomic Nervous System. 3nd ed. Amsterdam: Elsevier; 2012. pp. 3-8. DOI: 10.1016/B978-0-12-386525-0.00001-9

[3] Abel PW, Piascik MT. Introduction to autonomic nervous system drugs. In: Pharmacology and Therapeutics for Dentistry. Seventh ed. Elsevier; 2016. pp. 71-81

[4] Ganz J. Gut feelings: Studying enteric nervous system development, function, and disease in the zebrafish model system. Developmental Dynamics. 2018;**247**:268-278. DOI: 10.1002/DVDY.24597

[5] Wehrwein EA, Orer HS, Barman SM. Overview of the anatomy, physiology, and pharmacology of the autonomic nervous system. Comprehensive Physiology. 2016;**6**:1239-1278. DOI: 10.1002/cphy.c150037

[6] Fregoso SP, Hoover DB. Development of cardiac parasympathetic neurons, glial cells, and regional cholinergic innervation of the mouse heart. Neuroscience. 2012;**221**:28-36

[7] Kiyono K, Hayano J, Watanabe E, Yamamoto Y. Heart rate variability (HRV) and sympathetic nerve activity. In: Iwase S, Hayano J, Orimo S, editors. Clinical Assessment of the Autonomic Nervous System. First ed. Japan: Springer; 2017. pp. 147-161. DOI: 10.1007/978-4-431-56012-8_9

[8] Gerardo GM, Williams DP, Kessler M, et al. Body mass index and parasympathetic nervous system reactivity and recovery following graded exercise. American Journal of Human Biology. 2019;**31**:e23208. DOI: 10.1002/ajhb.23208

[9] Heart rate variability. Standards of measurement, physiological interpretation, and clinical use. Task Force of the European Society of Cardiology and the North American Society of Pacing and Electrophysiology. European Heart Journal. 1996;**17**: 354-381

[10] Bigger JT Jr, Fleiss JL, Steinman RC, Rolnitzky LM, Schneider WJ, Stein PK. RR variability in healthy, middle age persons compared with patients with chronic coronary heart disease or recent acute myocardial infarction. Circulation. 1995;**91**:1936-1943

[11] Vanderlei LC, Pastre CM, Hoshi RA, Carvalho TD, Godoy MF. Basic notions of heart rate variability and its clinical applicability. Revista Brasileira de Cirurgia Cardiovascular. 2009;**24**(2): 205-217. DOI: 10.1590/s0102-76382009000200018

[12] Shafer F, Ginsberg JP. An overview of heart rate variability metrics and norms. Frontiers in Public Health. 2017; **5**:258. DOI: 10.3389/ fpubh. 2017.00258

[13] Lombardi F, Huikuri H, Schmidt G, Malik M. Short-term heart rate variability: Easy to measure, difficult to interpret. On behalf of the e-rhythm study Group of European Heart Rhythm Association. Heart Rhythm. 2018; **15**(10):1559-1560. DOI: 10.1016 /j.hrthm.2018.05.023

[14] Goldberger JJ, Challapalli S, Tung R, Parker MA, Kadish AH. Relationship of heart rate variability to parasympathetic effect. Circulation. 2001;**1983**(103): 1977-1983

[15] Moher D, Liberati A, Tetzlaff J, Altman DG, The PRISMA Group.

Preferred reporting items for systematic reviews and meta-analyses: The PRISMA statement. PLoS Medicine. 2009;**6**(7):e1000097. DOI: 10.1371/journal.pmed1000097

[16] Bonnemeier H, Richardt G, Potratz J, Wiegand UK, Brandes A, Kluge N, et al. Circadian profile of cardiac autonomic nervous modulation in healthy subjects: Differing effects of aging and gender on heart rate variability. Journal of Cardiovascular Electrophysiology. 2003;**14**:791-799. DOI: 10.1046/j.1540-8167.2003.03078.x

[17] Almeida-Santos MA, Barreto-Filho JA, Oliveira JL, Reis FP, da Cunha Oliveira CC, Sousa AC. Aging, heart rate variability and patterns of autonomic regulation of the heart. Archives of Gerontology and Geriatrics. 2016;**63**: 1-8. DOI: 10.1016/j.archger.2015.11.011

Chapter 8

The Role of Magnetic Resonance Imaging (MRI) in Autonomic Nervous System Monitoring

Yousif Mohamed Y. Abdallah and Nouf H. Abuhadi

Abstract

Medical imaging of the nervous system is the methodology used to achieve pictures of parts of the nervous system for therapeutic uses to recognize the ailments. Magnetic resonance imaging (MRI) is a kind of medical imaging tool that utilizes solid magnetic fields and radio waves to deliver point-by-point pictures of the inside of the body. There are large number of imaging methodologies done each week around the world. Medical imaging is developing rapidly due to developments in image acquisition tools including functional MRI and hybrid imaging modalities. This chapter abridged the role of magnetic resonance imaging (MRI) in autonomic nervous system monitoring. This chapter also summarizes the image interpretation challenges in diagnosing autonomic nervous system disorders.

Keywords: medical, imaging, autonomic nervous system

1. Introduction

The nervous system is divided into two parts, the central (CNS) and peripheral (PNS) part. The CNS includes the majority of the neural tissues and comprises the brain and spinal cord. PNS comprises all the structures outside the CNS and includes the special sense, spinal and cranial, and autonomic nervous system (ANS) [1–4]. The nervous system is composed mostly of the axons of sensual and motor neurons that permit between the CNS and the body. The autonomic sensory system (ANS) is divided into the peripheral sensory parts that provision the muscles and organs and influence the capacity of inner organs [5–7]. This system is considered as a regulatory framework that stimulates the action of those organs and muscles. This system manages in essence capacities, for example, the pulse, absorption, optical reaction, pee, and voluptuous stimulation [8–11]. This framework is the essential instrument responsible for the battle or flight reaction. Inside the mind, the central nerves manage this system. Autonomic capacities incorporate control of breath, heart guideline (the cardiovascular control focus), vasomotor action (the vasomotor focus), and certain reflex activities, for example, hacking, wheezing, gulping, and heaving [11–14]. This system is then subdivided into different zones that are connected additionally to ANS and sensory structures outside to the cerebrum. The central nerve over the cerebrum trunk goes as an integrator for autonomic capacities, accepting ANS administrative contribution from the limbic framework to do as such. The ANS has three subdivisions: the thoughtful sensory, the parasympathetic sensory, and the enteric anxious system. [15–18] Some researchers exclude the enteric sensory as a component of this organization. [8] The thoughtful sensory organization frequently includes "fight or flight" framework, although the parasympathetic sensory organization regularly includes the "rest and digest" or "feed and breed" framework. Most of the time, both of these frameworks have "inverse" activities where one framework actuates a physiological reaction and the other hinders it [19–23]. A more established improvement of thoughtful and parasympathetic structures as "excitatory" and "inhibitory" was toppled because of the numerous exemptions found. In ANS, there are many constrainers and excitatory neurotransmitters, which locate among neural cells.

The non-noradrenergic system affects the gut and the lungs [24, 25]. Magnetic resonance imaging (MRI) is a medicinal imaging method utilized to frame photos of the life systems and the functional procedures of the body. MRI machines utilize solid magnetic fields and RF pulse to create pictures of the structures of the body. MRI does not use ionizing radiation like CT, PET, and other scanners. MRI is a restorative utilization of nuclear magnetic resonance (NMR) [26–28]. This technique can be utilized for NMR spectroscopy. Although the risks of conventional radiography are presently very much protected in utmost medicinal settings, an MRI examination may at present be viewed as a superior decision than a CT exam. MRI is generally utilized in emergency clinics and facilities for therapeutic determination. An MRI may produce diverse data in contrast to CT scan. There might be dangers and inconvenience related to MRI scans. In contrast to CT filters, this procedure commonly is more intense and risky. In the 1970–80s, MRI has demonstrated to be a flexible imaging method. Although MRI is utmost unmistakably utilized in analytic prescription and biological researches, it additionally might be utilized to make pictures of inorganic particles. The supported increment sought after for MRI inside wellbeing frameworks has prompted worries about cost adequacy and overdiagnosis [29–32].

2. Anatomy of autonomic nervous system (ANS)

The ANS is partitioned into the thoughtful and the parasympathetic sensory system. The thoughtful division starts in the thoracic spines and ends up in the L2–3. The parasympathetic division includes both cranial (III, IX, X) and sacral (S2–4) nerves (**Figure 1**) [33, 34].

The thoughtful sensory system consists of neural cells that appear beyond T1 and continue to level L2/3. There are a few areas whereupon preganglionic neurons can be able neurotransmitters because of their postganglionic neuron stability.

These ganglia assign the postganglionic neurons beside which innervation of goal structures pursues. Instances regarding splanchnic (instinctive) nerves are as follows:

1. Cervical cardiovascular nerves then thoracic instinctive nerves, which are neural ligature of the thoughtful band

2. Thoracic splanchnic nerves

3. Lumbar splanchnic nerves, which are neural connection of the prevertebral ganglia

4. Sacral splanchnic nerves, which are neural concretion of the second quantity hypogastric plexus [35–40]

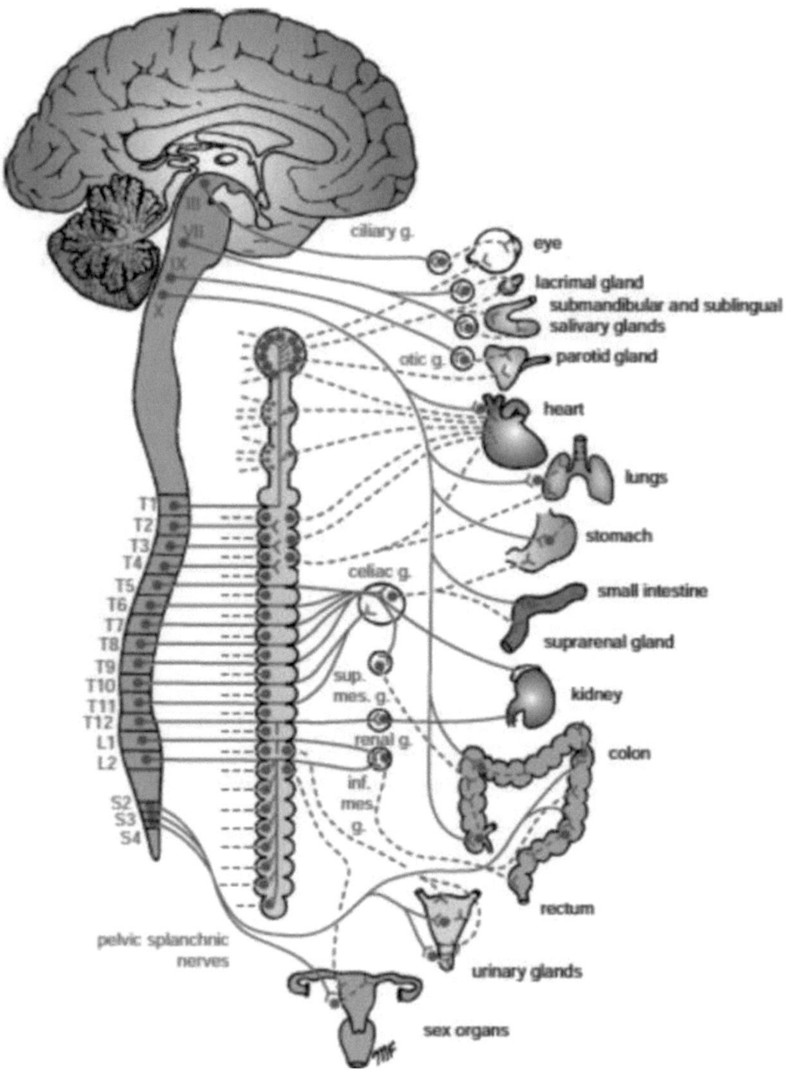

Figure 1.
Autonomic nervous system [1, 3, 8].

2.1 Sensory neurons

The sensory part is taken outdoors concerning necessary instinctive true neurons determined in the hem sensory dictation (PNS), of cranial real ganglia: the geniculate, petrosal, or nodose ganglia, annexed one at a time after cranial nerves. These tactile neurons are responsible of organization of the degrees of charcoal dioxide, oxygen, or grit between the blood, blood boat ounce yet the artificial business enterprise about the belly and intestine content [41–44]. The nTS gets the performance beside an adjacent chemosensory focus, the area postrema, who recognizes poisons among the blood yet the cerebrospinal melted and is necessary because synthetically instigated spewing and restrictive style repugnance (the intelligence as ensures so a life as has been harmed through sustenance in no way connection such again). These instinctive tactile data constantly then unknowingly regulate the labor regarding the machine neurons about the ANS (**Figure 2**) [45, 46].

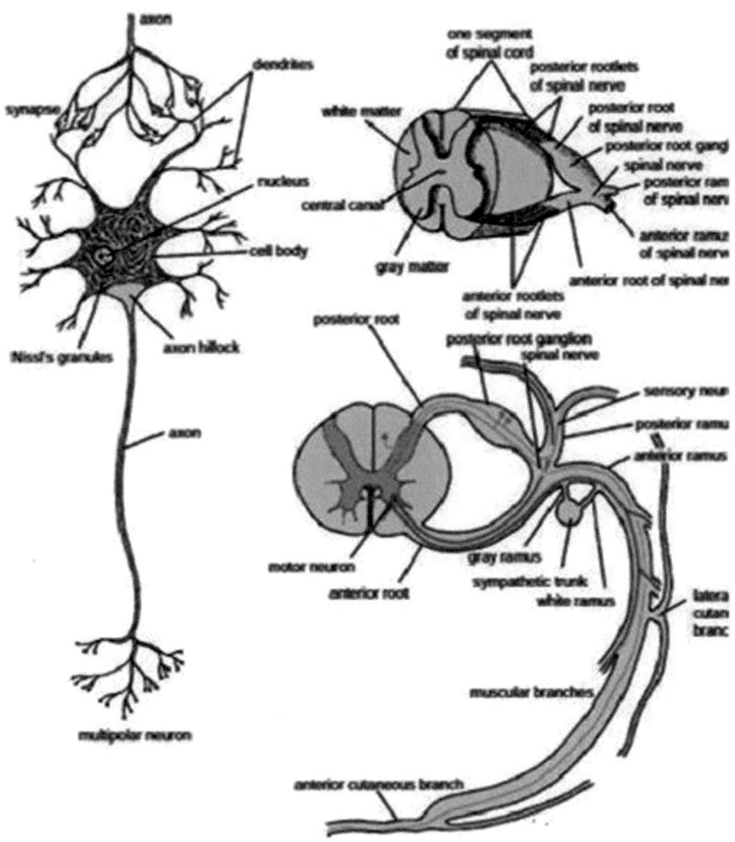

Figure 2.
Sensory neurons [1, 3, 8].

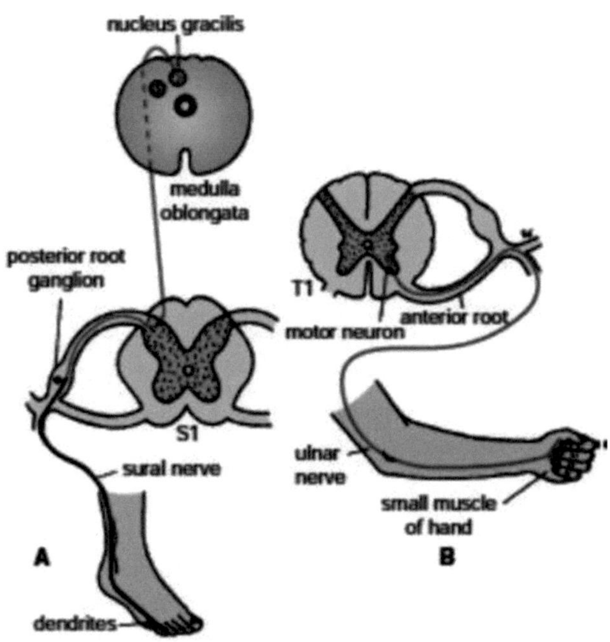

Figure 3.
The central and peripheral nervous system [1, 3, 8].

2.2 Innervation

Autonomic nerves travel in accordance with organs via the entire body. The true portion on the of the autonomic nerves remaining achieves the spinal piece at definitive spinal fragments. The neural signal travel from the autonomic system to the other body part through number pf the nerves that distributed throughout the body (**Figure 3**) [47, 48].

3. Physiology of autonomic nervous system

Thoughtful and parasympathetic divisions regularly work contrary to one another. Yet, this resistance is better named reciprocal in nature as opposed to hostile. The thoughtful partition regularly works in activities needing fast reactions. The thoughtful framework is regularly the "battle or flight" framework, while the other framework is frequently the "rest and summary" or "feed and breed" framework [49–51]. In any case, numerous cases of thoughtful and parasympathetic movement cannot be credited to "battle" or "rest" circumstances. For example, adjustable over out of a leaning again and placing role would contain an unsustainable decline between circulatory pressure notwithstanding a compensatory rise within the blood vessel's thoughtful tonus. Another mannequin is the steady, second-to-second tweak of the bough with the aid of thoughtful then parasympathetic impacts, so an aspect on the respiratory cycle. When all is said and done, these two frameworks ought to be viewed as for all time tweaking imperative capacities, in normally hostile design, to accomplish homeostasis. Higher living beings keep up their honesty by means of homeostasis, which depends on negative criticism guideline, which, thusly, ordinarily relies upon the autonomic anxious system [52–55]. Some run-of-the-mill activities of the thoughtful and parasympathetic sensory systems are recorded beneath [55].

4. Pathology of autonomic nervous system

4.1 Sweating abnormalities

Sudomotor or perspiring changes can likewise be highlights of autonomic brokenness, inferring changes in perspiring not related legitimately to side effects of orthostatic narrow mindedness or on the other hand presyncope [56–58]. Patients may report either expanded or on the other hand over the top perspiring or diminished perspiration yield and warmth narrow mindedness, either internationally, segmentally, or on the other hand sketchy in appropriation. Numerous patients with distal perspiration misfortune report expanded perspiration yield, which may happen as a compensatory reaction is unaffected territories, for example, the head and upper-middle, yet which is seen by the patient as unnecessary perspiring [59].

Sudomotor brokenness might be because of anomalies in focal control instruments (as in the different framework decay), or all the more generally in patients with autonomic fringe neuropathy, either as a disconnected variation from the norm of postganglionic thoughtful nerve strands just in hypohidrosis or worldwide anhidrosis, or as a component of an increasingly summed up autonomic neuropathy, either essential (immune system autonomic neuropathy) or auxiliary (amyloidosis, diabetic fringe neuropathy, or little fiber tangible neuropathy because of Sjögren's disorder) in nature [60, 61].

4.2 Secretomotor symptoms

Secretomotor indications incorporate sicca manifestations of dry eyes (xerophthalmia) and dry mouth (xerostomia). Patients do not visit the physicians for more investigations unless they becomes serious, however, with cautious addressing, they might be evoked. The brokenness of autonomic innervation might be seen in autonomic neuropathies or part of summed up autonomic disappointment, albeit even more ordinarily found previously [62–65].

5. Magnetic resonance imaging (MRI)

For MRI examination, the patient is situated inside an MRI scanner up to expectation constructions a consolidated alluring discipline around the sector in imitation of keep imaged. In utmost therapeutic applications, protons (hydrogen particles) that containing cloud particles was passed into tissues in order to create a sign that later use to make a photograph of internal structure of the body. Initially, energy of swaying magnetic field is temporarily related after the patient at the becoming reverberation recurrence. The energized hydrogen iotas beam a radio recurrence signal, which is estimated with the aid of an accepting curl. The radio sign may stay instituted to encode role data with the aid of altering the foremost pleasing subject utilizing bias loops. As those curls are rapidly became concerning or far away that redact the trademark stupid concussion on an MRI check. The difference in a number of tissues is managed by using the dimensions at which energized particles appear returned to a coherent state. Exogenous division specialists would possibly lie fond in accordance with the unaccompanied in conformity to perform the photograph clearer. [65] The actual parts of an MRI machine are precept magnet and the RF framework, which admits the NMR signal. The complete framework is restrained by using at least certain PCs. The area virtue on the magnet is estimated in teslas then preserving in thinking so just concerning the frameworks labor at 1.5 T, business frameworks are on hand someplace in the extent concerning 0.2 yet 7 T. For claustrophobic patient usually the open superconducting magnet machine is used. Recently, MRI has been shown either at ultra-low fields. The place ample sign quality is done conceivable via prepolarization (on the pray of 10 up to −100 mT) then by estimating the Larmor antecedence fields at around one hundred microteslas including very delicate superconducting quantum arrest gadgets (SQUIDs) [66]. Each art comes lower back according to its harmony administration and then exasperation by using the unrestricted unwinding approaches regarding T1 or T2. The T1 weighted picture is treasured because surveying the brain tissues, distinguishing greasy structure, describing average lung accidents and now every is pointed out in performed because

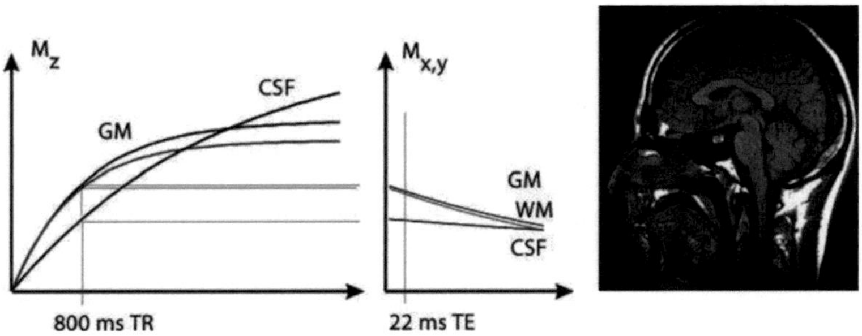

Figure 4.
MRI T1-weighted image [30, 64–66]

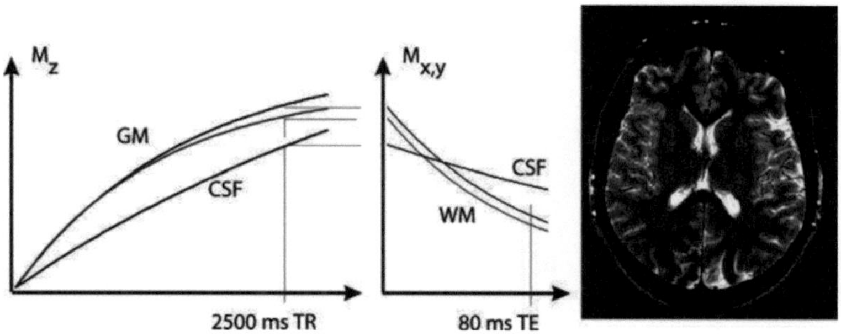

Figure 5.
MRI T2-weighted image [30, 64–66].

of acquiring morphological data, simply namely because of post-differentiate imaging (**Figure 4**) [30, 67, 68].

The T2-weighted picture shows a valuable structure for identifying and recognition of the pathophysiological problems of ANS and gives useful information that enable the neurosurgeons to perform a successful procedure (**Figure 5**).

6. Conclusion

In nervous system disorders, the non-ionizing radiation is used to scan and produce multi-dimension images with and without contrast media utilization. In the 1970s, Ian Robert Young and Hugh Clow had first brain images using MRI. In 1990, Seiji Ogawa who used the oxygen-depleted blood phenomena introduced functional MRI (fMRI). In the 1990s, the development and introduction of the new MRI protocols helped in the demonstration of gray and white matter areas of the nervous system. Many MRI scans later were done by using high magnetic strength (3.0 up to 9.4 T).

Acknowledgements

The authors are thankful to the Deanship of Scientific Research, at Majmaah University, for funding this research.

Conflict of interest

There are no conflicts of interest.

Author details

Yousif Mohamed Y. Abdallah[1*] and Nouf H. Abuhadi[2]

1 Radiological Science and Medical Imaging Department, College of Applied Medical Science, Majmaah University, Majmaah, Saudi Arabia

2 Diagnostic Radiology Department, College of Applied Medical Science, Jazan University, Jazan, Saudi Arabia

*Address all correspondence to: y.yousif@mu.edu.sa

IntechOpen

© 2019 The Author(s). Licensee IntechOpen. This chapter is distributed under the terms of the Creative Commons Attribution License (http://creativecommons.org/licenses/by/3.0), which permits unrestricted use, distribution, and reproduction in any medium, provided the original work is properly cited. (cc) BY

References

[1] Schmidt A, Thews G. Autonomic nervous system. In: Janig W, editor. Human Physiology. 2nd ed. New York, NY: Springer-Verlag; 1989. pp. 333-370

[2] Allostatic Load Notebook: Parasympathetic Function Archived 2012-08-19 at the Wayback Machine - 1999, MacArthur research network, UCSF

[3] Langley JN. The Autonomic Nervous System Part 1. Cambridge: W. Heffer; 1921

[4] Jänig W. Integrative Action of the Autonomic Nervous System: Neurobiology of Homeostasis (Digitally Printed Version). Cambridge: Cambridge University Press; 2008. p. 13. ISBN 978052106754-6

[5] Abdallah Y. Improvement of sonographic appearance using HAT-TOP methods. International Journal of Science and Research (IJSR). 2015;**4**(2):2425-2430

[6] John F. Enteric Nervous System. Scholarpedia. Archived from the original on 8 October 2017; 2007. Retrieved 8 October 2017. DOI: 10.4249/scholarpedia.4064

[7] Willis WD. The autonomic nervous system and its central control. In: Berne RM, editor. Physiology. 5th ed. St. Louis, Mo: Mosby; 2004. ISBN 0323022251

[8] Pocock G. Human Physiology. 3rd ed. London, United Kingdom: Oxford University Press; 2006. pp. 63-64. ISBN 978-0-19-856878-0

[9] Belvisi Maria G, David Stretton C, Yacoub Magdi, Barnes Peter J. "Nitric oxide is the endogenous neurotransmitter of bronchodilator nerves in humans". European Journal of Pharmacology. 1992; **210**(2):221-222. DOI: 10.1016/0014-2999(92)90676-U. PMID 1350993

[10] Costanzo LS. Physiology. Hagerstwon, MD: Lippincott Williams & Wilkins; 2007. p. 37. ISBN 0-7817-7311-3

[11] Moore KL, Agur AM. Essential Clinical Anatomy. 2nd ed. Lippincott Williams & Wilkins, Inc.; 2002. p. 199. ISBN 978-0-7817-5940-3

[12] Neil A. Campbell, Jane B. Reece: Biologie. Spektrum-Verlag Heidelberg-Berlin; 2003. ISBN 3-8274-1352-4

[13] Goldstein D. Principles of Autonomic Medicine (PDF) (Free Online Version Ed.). Bethesda, Maryland: National Institute of Neurological Disorders and Stroke, National Institutes of Health, USA; 2016. ISBN 9780824704087

[14] Hadhazy A. Think Twice: How the Gut's "Second Brain" Influences Mood and Well-Being. Scientific American; 2010. Archived from the original on December 31, 2017

[15] Zimmerman-Viehoff F, Thayer J, Koenig J, Herrmann C, Weber CS, Deter H-C. Short-term effects of espresso coffee on heart rate variability and blood pressure in habitual and non-habitual coffee consumers- a randomized crossover study. Nutritional Neuroscience. 2016;**19**(4):169-175. Retrieved February 20, 2017

[16] Bunsawat K, White DW, Kappus RM, Baynard T. Caffeine delays autonomic recovery following acute exercise. European Journal of Preventive Cardiology. 2015;**22**(11):1473-1479. Retrieved February 20, 2017

[17] Monda M, Viggiano An, Vicidomini C, Viggiano Al, Iannaccone T, Tafuri D, De Luca,

B. "Espresso coffee increases parasympathetic activity in young, healthy people". Nutritional Neuroscience. 2009;**12**(1):43-48. Retrieved February 20, 2017

[18] Information Reed Business. Britain's Brains Produce First NMR Scans. New Scientist: 588;1978

[19] Blood-flow checker. Popular Science: 12;1987

[20] Le Bihan D, Breton E. Method to Measure the Molecular Diffusion and/or Perfusion Parameters of Live Tissue. US Patent # 4809701;1987

[21] Abdallah YM. History of medical imaging. Archives of Medicine and Health Sciences. 2017;**5**:275-278

[22] Villringer A, Rosen BR, Belliveau JW, Ackerman JL, Lauffer RB, Buxton RB, et al. Dynamic imaging with lanthanide chelates in normal brain: Contrast due to magnetic susceptibility effects. USA: Magnetic Resonance in Medicine. 1988;**6**(2):164-174

[23] Faro SH, Mohamed FB. Bold fMRI. A Guide to Functional Imaging for Neuroscientists. Springer; 2010. ISBN 978-1-4419-1328-9. Retrieved 10 June 2015

[24] Howe FA, Filler AG, Bell BA, Griffiths JR. Magnetic resonance neurography. Magnetic Resonance in Medicine. 1992;**28**(2):328-338. DOI: 10.1002/mrm.1910280215. PMID: 1461131

[25] Filler AG, Howe FA, Hayes CE, Kliot M, Winn HR, Bell BA, et al. Magnetic resonance neurography. Lancet. 1993;**341**(8846):659-661. DOI: 10.1016/0140-6736(93)90422-d. PMID 8095572

[26] Filler A. Magnetic resonance neurography and diffusion tensor imaging: Origins, history, and clinical impact of the first 50,000 cases with an assessment of efficacy and utility in a prospective 5000-patient study group. Neurosurgery. USA. 2009;**65** (4 Suppl):A29-A43. DOI: 10.1227/01.neu.0000351279.78110.00. PMC 2924821. PMID 19927075

[27] Basser PJ. Invention and Development of Diffusion Tensor MRI (DT-MRI or DTI) at the NIH. Diffusion MRI. Oxford University Press; 2010. pp. 730-740. DOI: 10.1093/med/9780195369779.003.0047. ISBN 9780195369779

[28] Hajnal JV, De Coene B, Lewis PD, Baudouin CJ, Cowan FM, Pennock JM, et al. High signal regions in normal white matter shown by heavily T2-weighted CSF nulled IR sequences. Journal of Computer Assisted Tomography. 1992;**16**(4):506-513. DOI: 10.1097/00004728-199207000-00002. PMID: 1629405

[29] Koretsky AP. Early development of arterial spin labeling to measure regional brain blood flow by MRI. NeuroImage. 2012;**62**(2):602-607. DOI: 10.1016/j.neuroimage.2012.01.005. PMC 4199083. PMID 22245338

[30] Abdallah Y. An Introduction to PACS in Radiology Service: Theory and Practice. Vol. 2. Berlin, Germany: LAP LAMBERT Academic Publishing; 2012. pp. 140-153. ISBN 978-3846588987

[31] Reichenbach JR, Venkatesan R, Schillinger DJ, Kido DK, Haacke EM. Small vessels in the human brain: MR venography with deoxyhemoglobin as an intrinsic contrast agent. Radiology. July 1997;**204**(1):272-277. DOI: 10.1148/radiology.204.1.9205259. PMID: 9205259

[32] Mansfield P, Coxon R, Glover P. Echo-planar imaging of the brain at 3.0 T: First normal volunteer results. Journal of Computer Assisted

Tomography. 1994;**18**(3):339-343. DOI: 10.1097/00004728-199405000-00001. PMID 8188896

[33] Robitaille PM, Abduljalil AM, Kangarlu A, Zhang X, Yu Y, Burgess R, et al. Human magnetic resonance imaging at 8 T. NMR in Biomedicine. 1998;**11**(6):263-265. DOI: 10.1002/(SICI)1099-1492(199810)11:6<263:AID-NBM549>3.0.CO;2-0. PMID: 9802467

[34] Abdallah Y. Increasing of edges recognition in cardiac Scintigraphy for ischemic patients. Journal of Biomedical Engineering and Medical Imaging. 2016;**2**(6):39-40

[35] Vaughan T, DelaBarre L, Snyder C, Tian J, Akgun C, Shrivastava D, et al. 9.4T human MRI: Preliminary results. USA: Magnetic Resonance in Medicine. 2006;**56**(6):1274-1282. DOI: 10.1002/mrm.21073. PMC 4406343. PMID 17075852

[36] Ebel K, Benz-Bohm G. Differential Diagnosis in Pediatric Radiology. Thieme; 1999. 538 p. ISBN 978-3-13-108131-5. Retrieved 18 July 2011

[37] Bradley WG, Brant-Zawadzki M, Cambray-Forker J. MRI of the Brain. India: Surendra Kumar; 2001. ISBN 978-0-7817-2568-2. Retrieved 24 July 2011

[38] Roos KL, Tunkel AR. Bacterial Infections of the Central Nervous System. India: Elsevier Health Sciences; 2010. 69 p. ISBN 978-0-444-52015-9. Retrieved 18 July 2011

[39] American Medical Society for Sports Medicine. "Five Things Physicians and Patients Should Question," Choosing Wisely: An initiative of the ABIM Foundation, American Medical Society for Sports Medicine; 2014. Retrieved 29 July 2014

[40] Churchill Nathan W, Hutchison Michael G, Doug R, General L, Graham Simon J, Schweizer Tom A. The first week after concussion: Blood flow, brain function and white matter microstructure. NeuroImage: Clinical. USA. 2017;**14**:480-489

[41] Garel C. MRI of the Fetal Brain: Normal Development and Cerebral Pathologies. USA: Springer; 2004. ISBN 978-3-540-40747-8. Retrieved 24 July 2011

[42] Butler P, Mitchell AW, Ellis H. Applied Radiological Anatomy for Medical Students. United Kingdom: Cambridge University Press; 2007. 12 p. ISBN 978-0-521-81939-8. Retrieved 18 July 2011

[43] Tofts P. Quantitative MRI of the Brain: Measuring Changes Caused by Disease. USA: John Wiley and Sons; 2005. 86 p. ISBN 978-0-470-86949-9. Retrieved 18 July 2011

[44] Chowdhury R, Wilson I, Rofe C, Lloyd-Jones G. Radiology at a Glance. USA: John Wiley and Sons; 2010. 95 p. ISBN 978-1-4051-9220-0. Retrieved 18 July 2011

[45] Granacher RP. Traumatic Brain Injury: Methods for Clinical and Forensic Neuropsychiatric Assessment. CRC Press. 247 p; 2007. ISBN 978-0-8493-8138-6. Retrieved 18 July 2011

[46] Goss CH, Ratjen F. Update in cystic fibrosis. American Journal of Respiratory and Critical Care Medicine 2013. 2012;**187**:915-919. DOI: 10.1164/rccm.201301-0184UP

[47] Elborn JS. Cystic fibrosis. The Lancet. 2016;**388**:2519-2531. DOI: 10.1016/S0140-6736(16)00576-6

[48] Ramsey BW, Welsh MJAJRCCM. 100-year anniversary. Progress along the pathway of discovery leading to treatment and cure of cystic fibrosis. American Journal of Respiratory and Critical Care Medicine. 2017;**195**:1092-1099

[49] Ratjen F, Bell SC, Rowe SM, et al. Cystic fibrosis. Nature Reviews. Disease Primers. 2015;**1**

[50] Ong T, Ramsey BW. Update in cystic fibrosis. American Journal of Respiratory and Critical Care Medicine 2015. 2014;**192**:669-675. DOI: 10.1164/rccm.201504-0656UP

[51] Weill D, Benden C, Corris PA, et al. A consensus document for the selection of lung transplant candidates: 2014--an update from the pulmonary transplantation Council of the International Society for heart and lung transplantation. The Journal of Heart and Lung Transplantation. 2015;**34**:1-15

[52] Reznikov LR. Cystic fibrosis and the nervous system. Chest. 2017;**151**:1147-1155

[53] Liou TG. The clinical biology of cystic fibrosis transmembrane regulator protein: Its role and function in extrapulmonary disease. Chest. 2019;**155**:605-616

[54] Willison HJ, Muller DP, Matthews S, et al. A study of the relationship between neurological function and serum vitamin E concentrations in patients with cystic fibrosis. Journal of Neurology, Neurosurgery, and Psychiatry. 1985;**48**:1097-1102

[55] Cynamon HA, Milov DE, Valenstein E, et al. Effect of vitamin E deficiency on neurologic function in patients with cystic fibrosis. The Journal of Pediatrics. 1988;**113**:637-640

[56] Hamid B, Khan A. Cerebral hemorrhage as the initial manifestation of cystic fibrosis. Journal of Child Neurology. 2007;**22**:114-5. 12-46

[57] Mérelle ME, Griffioen RW, Dankert-Roelse JE. Cystic fibrosis presenting with intracerebral haemorrhage. Lancet. 2001;**358**

[58] Sritippayawan S, MacLaughlin EF, Woo MS. Acute neurological deficits in a young adult with cystic fibrosis. Pediatric Pulmonology. 2003;**35**:147-5114

[59] Rao DS, Infeld MD, Stern RC, et al. Cough-induced hemiplegic migraine with impaired consciousness in cystic fibrosis. Pediatric Pulmonology. 2006;**41**:171-176

[60] Goldstein AB, Goldstein LS, Perl MK, et al. Cystic fibrosis patients with and without central nervous system complications following lung transplantation. Pediatric Pulmonology. 2000;**30**:203-206

[61] Živković SA, Jumaa M, Barišić N, et al. Neurologic complications following lung transplantation. Journal of the Neurological Sciences. 2009;**280**:90-93

[62] Atlante A, Favia M, Bobba A, et al. Characterization of mitochondrial function in cells with impaired cystic fibrosis transmembrane conductance regulator (CFTR) function. Journal of Bioenergetics and Biomembranes. 2016;**48**:197-210

[63] Kelly M, Trudel S, Brouillard F, et al. Cystic fibrosis transmembrane regulator inhibitors CFTR(inh)-172 and GlyH-101 target mitochondrial functions, independently of chloride channel inhibition. The Journal of Pharmacology and Experimental Therapeutics. 2010;**333**:60-69

[64] Abdallah Y. Application of Analysis Approach in Noise Estimation, Using Image Processing Program. Germany: Lambert Publishing Press GmbH & Co. KG; 2011. pp. 123-125 [4]

[65] Abdallah Y, Yousef R. Augmentation of X-rays images using pixel intensity values adjustments. International Journal of Science and Research (IJSR). 2015;**4**(2):2425-2430

[66] Abdallah Y. Increasing of Edges Recognition in Cardiac Scintography for Ischemic Patients. Germany: Lambert Publishing Press GmbH & Co. KG; 2011. pp. 123-125

[67] Abdallah YMY, Abdallah M. Using basic morphology tools in improvement of kidneys detection. International Journal of Science and Research (IJSR). 2015;**4**(5):1383-1386

[68] Abdallah Y. Increasing of edges recognition in cardiac scintigraphy for ischemic patients. Journal of Biomedical Engineering and Medical Imaging. 2016;**2**(6):40-48